SpringerBriefs in Mathematics

SpringerBriefs in Mathematics showcases expositions in all areas of mathematics and applied mathematics. Manuscripts presenting new results or a single new result in a classical field, new field, or an emerging topic, applications, or bridges between new results and already published works, are encouraged. The series is intended for mathematicians and applied mathematicians.

More information about this series at http://www.springer.com/series/10030

Michael Grabchak

Tempered Stable Distributions

Stochastic Models for Multiscale Processes

Springer

Michael Grabchak
Department of Mathematics and Statistics
University of North Carolina
Charlotte, NC, USA

ISSN 2191-8198 ISSN 2191-8201 (electronic)
SpringerBriefs in Mathematics
ISBN 978-3-319-24925-4 ISBN 978-3-319-24927-8 (eBook)
DOI 10.1007/978-3-319-24927-8

Library of Congress Control Number: 2015957382

Mathematics Subject Classification (2010): 60E07, 60G51, 60G52

Springer Cham Heidelberg New York Dordrecht London

Printed on acid-free paper

Springer International Publishing AG Switzerland is part of Springer Science+Business Media (www.springer.com)

To Lijuan

Acknowledgments

I would like to thank Professor Gennady Samorodnitsky, my PhD advisor, who first introduced me to tempered stable distributions and spent many hours discussing them with me. I likewise thank Donna Chernyk, editor at Springer, who first suggested the possibility of me writing this brief and then helped me through the entire process. I also thank the many anonymous referees who read the manuscript both while it was being written and after the first draft was completed. Their valuable comments and suggestions led to improvements in the presentation of this brief. Finally, I thank my family for their support and patience.

Contents

Notation

Let $\mathbb{R}^d$ be the space of d-dimensional column vectors of real numbers. For $x \in \mathbb{R}^d$ we write $x = (x_1, x_2, \ldots, x_d)$, and we denote the transpose of x by x^T. Let $\langle \cdot, \cdot \rangle$ be the usual inner product on $\mathbb{R}^d$, i.e., if $x, y \in \mathbb{R}^d$ then $\langle x, y \rangle = x^T y = \sum_{i=1}^d x_i y_i$. Let $|\cdot|$ be the usual norm on $\mathbb{R}^d$, i.e., if $x \in \mathbb{R}^d$ then $|x| = \sqrt{\langle x, x \rangle} = \sqrt{\sum_{i=1}^d x_i^2}$. Let $\mathbb{S}^{d-1} = \{x \in \mathbb{R}^d : |x| = 1\}$, and let $\mathbb{C}^d$ be the space of d-dimensional column vectors of complex numbers. For $z \in \mathbb{C}^d$ we write $\Re z$ to denote the real part of z and $\Im z$ to denote the imaginary part of z. Let $\mathbb{R}^{d \times d}$ be the collection of all $d \times d$-dimensional matrices with real entries, and for $A \in \mathbb{R}^{d \times d}$ let $\mathrm{tr} A$ be the trace of A. Let $\mathfrak{B}(\mathbb{R}^d)$ and $\mathfrak{B}(\mathbb{S}^{d-1})$ denote the Borel sets on $\mathbb{R}^d$ and $\mathbb{S}^{d-1}$, respectively.

If μ is a probability measure on $(\mathbb{R}^d, \mathfrak{B}(\mathbb{R}^d))$, we write $\hat{\mu}(z) = \int_{\mathbb{R}^d} e^{i\langle x, z\rangle} \mu(\mathrm{d}x)$ for $z \in \mathbb{R}^d$ to denote its characteristic function, and we write $X \sim \mu$ to denote that X is a random variable taking values in $\mathbb{R}^d$ with distribution μ. Further, we write $X_1, X_2, \ldots, X_n \stackrel{\text{iid}}{\sim} \mu$ to denote that $X_1, X_2, \ldots, X_n$ are independent and identically distributed random variables taking values in $\mathbb{R}^d$, each with distribution μ. For two random variables X and Y, we write $X \stackrel{d}{=} Y$ to denote that X and Y have the same distribution.

For a sequence of random variables $X_1, X_2, \ldots$, we write $\text{d-lim} X_n$ to denote the limit in distribution. We also use the more standard notations $\stackrel{d}{\to}$ and $\stackrel{p}{\to}$ to denote, respectively, convergence in distribution and convergence in probability. For sequences of probability measures, we write $\stackrel{w}{\to}$ to denote weak convergence, and for sequences of Radon measures, we write $\stackrel{v}{\to}$ to denote vague convergence.

We write $ID(A, M, b)$ to denote the infinitely divisible distribution on $\mathbb{R}^d$ with Gaussian part A, Lévy measure M, and shift b. We write $N(b, A)$ to denote the Gaussian distribution on $\mathbb{R}^d$ with mean vector b and covariance matrix A, and we write $S_\alpha(\sigma, b)$ to denote the infinite variance α-stable distribution on $\mathbb{R}^d$ with spectral measure σ and shift b. We write $TS_\alpha^p(R, b)$ to denote the p-tempered α-stable distribution on $\mathbb{R}^d$ with Rosiński measure R and shift b, and we write $ETS_\alpha^p(A, \nu, b)$ to denote the extended p-tempered α-stable distribution on $\mathbb{R}^d$ with Gaussian part A, extended Rosiński measure ν, and shift b. Several parametric families of tempered stable distributions are introduced in Chapter 6. These are denoted by $STLF^p_\alpha(c_-, c_+, \ell_-, \ell_+, b)$, $TW^p_\alpha(c, \ell, b)$, $TW_\alpha(c, \ell, b)$, $PT^p_\alpha(c_-, c_+, \nu_-, \nu_+, b)$, and $GT^p_\alpha(c_-, c_+, \nu_-, \nu_+, \ell_-, \ell_+, b)$.

If $\mathscr{A}$ is a collection of subsets of some space $\mathbb{E}$, we write $\sigma(\mathscr{A})$ to denote the σ-algebra generated by $\mathscr{A}$, i.e., the smallest σ-algebra that contains $\mathscr{A}$. If f and g are real-valued functions, $c \in \{0, \infty\}$, and $k \in \mathbb{R}$, we write

$$f(t) \sim kg(t) \text{ as } t \to c$$

to denote

$$f(t)/g(t) \to k \text{ as } t \to c.$$

Note that this is somewhat nonstandard notation in the case when $k = 0$. When dealing with infinity we adopt the conventions that

$$1/\infty = 0 \text{ and } 1/0 = \infty.$$

For $x, y \in \mathbb{R}$ we write $x \wedge y$ to denote the minimum and $x \vee y$ to denote the maximum. Further, we define $\log^+ x := \log(1 \vee x)$. For a set H we denote the indicator function on H by 1_H. This means that $1_H(x) = 1$ if $x \in H$ and $1_H(x) = 0$ if $x \notin H$. We also write $1_{x \in H}$ to denote this. For $a \in \mathbb{R}^d$ we write δ_a to denote the Dirac delta measure at a. This means that for any Borel set H, we have $\delta_a(H) = 1_H(a)$. We denote the gamma function by $\Gamma(x)$. When $x > 0$ we have

$$\Gamma(x) = \int_0^\infty e^{-t} t^{x-1} \mathrm{d}t.$$

Further, we extend the gamma function to any $x \in \mathbb{R}$ with $x \neq 0, -1, -2, \ldots$ by the relation

$$\Gamma(x+1) = x\Gamma(x).$$

If P is a probability measure on the measurable space $(\Omega, \mathscr{F})$, we write E_P to denote the expectation with respect to P. Further, when there is no chance of ambiguity, we write E instead of E_P.

Chapter 1
Introduction

It has been observed that infinite variance stable distributions provide a good fit to data in a variety of situations. However, the extremely heavy tails of these models are not realistic for most real-world applications. In practice, there are all kinds of obstacles limiting the size of random phenomena. This has led researchers to use models that are similar to stable distributions in some central region, but with lighter tails. Tempered stable distributions are a rich class of models that capture this type of behavior. They have been shown to provide a good fit to data in a number of applications including actuarial science [36, 37], biostatistics [1, 39, 59], computer science [16, 74, 75], mathematical finance [17, 46, 50, 60], and physics [14, 55]. Discussions of the mechanisms by which such models appear in applications are given in [19, 35], and [34], see also Chapter 7 in this brief.

The idea of using models that are "stable-like" in the center but with lighter tails seems to have originated in the physics literature with the influential paper of Mantegna and Stanley [52]. That paper introduced the class of truncated Lévy flights.[1] These models start with a stable distribution and then truncate its tails. More formally, let $f(x)$ be the density of a one-dimensional infinite variance α-stable distribution and let $T > 0$ be a truncation level. We define a **truncated Lévy flight (TLF)** with truncation level T to be a probability measure with density

$$f_T(x) = c_T f(x) 1_{|x| \le T},$$

where c_T is a normalizing constant. If T is very large, then the models corresponding to the densities $f(x)$ and $f_T(x)$ may be statistically indistinguishable even for very

[1] Lévy flights are random walks, where the step sizes are independent and identically distributed draws from an infinite variance α-stable distribution, see [52]. The term "truncated Lévy flight" was also originally used to refer to a random walk, but has now come to refer to the underlying distribution.

M. Grabchak, *Tempered Stable Distributions*, SpringerBriefs in Mathematics, DOI 10.1007/978-3-319-24927-8_1

large datasets. However, the tail behavior of these models is vastly different. Stable distributions have an infinite variance, while TLFs have all moments finite.

Although these models capture the basic idea of modifying the tails of stable distributions to make them lighter, they have a number of limitations. First, unlike stable distributions TLFs are not infinitely divisible,[2] which means that the rich theory of infinitely divisible distributions cannot be used to analyze their behavior and that we cannot define Lévy processes with such marginal distributions. Another issue, one which has major ramifications for risk estimation, is that it is essentially impossible to estimate the truncation parameter T from a finite dataset, yet different values of T give vastly different estimates of risk. Aside from these considerations, it is generally desirable to allow for more flexible tail behavior than what is allowed by these models.

With issues such as these in mind, Koponen [48] suggested a different approach to modifying the tails of stable distributions to make them lighter. The idea begins by observing that an infinite variance α-stable distribution is infinitely divisible with no Gaussian part and a Lévy measure[3] given by

$$L(\mathrm{d}x) = c_-|x|^{-1-\alpha}1_{x<0}\mathrm{d}x + c_+x^{-1-\alpha}1_{x>0}\mathrm{d}x,$$

where $c_-, c_+ \geq 0$. Noting that the tails of the Lévy measure are intimately related to the tails of the distribution, the idea is to modify the tails of the Lévy measure to make them lighter and yet to keep the Lévy measure virtually unchanged in some central region. For this reason Koponen [48] introduced an infinitely divisible distribution with a Lévy measure given by

$$M(\mathrm{d}x) = c_-|x|^{-1-\alpha}e^{-|x|/\ell_-}1_{x<0}\mathrm{d}x + c_+x^{-1-\alpha}e^{-x/\ell_+}1_{x>0}\mathrm{d}x, \tag{1.1}$$

where $c_-, c_+, \geq 0$ and $\ell_-, \ell_+ > 0$. Clearly, if ℓ_- and ℓ_+ are very large, then the Lévy measure will be close to that of the corresponding α-stable distribution in the center and, potentially, quite far into the tails. However, ultimately, the tails of such a Lévy measure will decay exponentially fast. This leads to exponential decay in the tails of the corresponding probability measure as well.

Infinitely divisible distributions with no Gaussian part and a Lévy measure given by (1.1) have come to be known as **smoothly truncated Lévy flights (STLF)**. However, in the financial literature they are also sometimes referred to as the CGMY model (named after Carr, Geman, Madan, and Yor, the authors of [17]) or the KoBoL model (named after Koponen, the author of [48], and Boyarchenko and Levendorskiĭ, the authors of [15]). It should be mentioned that a number of special classes of STLFs had previously appeared in the literature. These include the inverse Gaussian (see, e.g., [71]) and some of its extensions given in [1, 39, 77], and [58].

[2]This is because infinitely divisible distributions cannot have a bounded support, see Theorem 24.3 in [69].

[3]Background on infinitely divisible distributions and Lévy measures is given in Section 2.2.

However, in these cases, the models were introduced not from the perspective of modifying the tails of a stable distribution, but from other considerations.

Despite the usefulness of STLFs, they have a number of limitations. For one thing they have exponential tails, which are too restrictive for many applications. In particular, there is evidence that the distributions of financial returns have regularly varying tails (see, e.g., [21]), yet this type of behavior cannot be captured by STLFs. Further, the class of STLFs is not closed under aggregation, i.e. the sum of two independent STLFs is not, in general, an STLF.

To deal with these limitations Rosiński [65] introduced the class of tempered α-stable distributions. The idea comes from considering the sum of n STLFs. Let $X_1, \ldots, X_n$ be independent random variables such that the distribution of X_i is that of an STLF with Lévy measure given by (1.1) and parameters $c^i_-, c^i_+, \ell^i_-, \ell^i_+$. We assume that all of the X_is have the same parameter α. The distribution of the sum $\sum_{i=1}^n X_i$ is infinitely divisible with no Gaussian part and a Lévy measure given by

$$\begin{aligned} M(\mathrm{d}x) &= |x|^{-1-\alpha}\left(\sum_{i=1}^n c^i_- e^{-|x|/\ell^i_-}\right) 1_{x<0}\mathrm{d}x + x^{-1-\alpha}\left(\sum_{i=1}^n c^i_+ e^{-x/\ell^i_+}\right) 1_{x>0}\mathrm{d}x \\ &= |x|^{-1-\alpha}\int_0^\infty e^{-|x|t} Q_-(\mathrm{d}t) 1_{x<0}\mathrm{d}x + x^{-1-\alpha}\int_0^\infty e^{-xt} Q_+(\mathrm{d}t) 1_{x>0}\mathrm{d}x, \end{aligned}$$

where $Q_-(\mathrm{d}t) = \sum_{i=1}^n c^i_- \delta_{1/\ell^i_-}(\mathrm{d}t)$ and $Q_+(\mathrm{d}t) = \sum_{i=1}^n c^i_+ \delta_{1/\ell^i_+}(\mathrm{d}t)$. This can easily be generalized by considering other forms for the measures $Q_-(\mathrm{d}t)$ and $Q_+(\mathrm{d}t)$. Further, this approach can be extended to d-dimensions by allowing a different measure $Q_u(\mathrm{d}t)$ for each direction $u \in \mathbb{S}^{d-1}$, where $\mathbb{S}^{d-1}$ is the unit sphere in d-dimensions. Specifically, noting that the Lévy measure of a d-dimensional α-stable distribution is given by

$$L(B) = \int_{\mathbb{S}^{d-1}}\int_0^\infty 1_B(ru) r^{-1-\alpha}\mathrm{d}r\sigma(\mathrm{d}u), \quad B \in \mathfrak{B}(\mathbb{R}^d),$$

where σ is a finite measure on $\mathbb{S}^{d-1}$, Rosiński [65] defined the class of **tempered α-stable distributions** as infinitely divisible distributions with no Gaussian part and a Lévy measure of the form

$$M(B) = \int_{\mathbb{S}^{d-1}}\int_0^\infty 1_B(ru) q(r,u) r^{-1-\alpha}\mathrm{d}r\sigma(\mathrm{d}u), \quad B \in \mathfrak{B}(\mathbb{R}^d), \tag{1.2}$$

where the function q, called **the tempering function**, is assumed to be of the form

$$q(r,u) = \int_0^\infty e^{-rt} Q_u(\mathrm{d}t)$$

for some measurable family of Borel measures $\{Q_u\}_{u \in \mathbb{S}^{d-1}}$.

In [27], the wider class of **p-tempered α-stable distributions** (TS^p_α), where $p > 0$ and $\alpha \in (-\infty, 2)$, was introduced. Here the tempering function is assumed to be of the form

$$\int_0^\infty e^{-r^p t} Q_u(\mathrm{d}t).$$

The parameter p controls the amount of tempering, while α is the index of stability of the corresponding stable distribution. Clearly, the case where $\alpha \le 0$ no longer has any meaning in terms of tempering stable distributions, however it allows the class to be more flexible. In fact, within certain subclasses, the case where $\alpha \le 0$ has been shown to provide a good fit to data, see, e.g., [1] or [17]. Rosiński's class corresponds to the case when $p = 1$ and $\alpha \in (0, 2)$. Tempered infinitely divisible distributions defined in [9] are another subclass corresponding to the case when $p = 2$ and $\alpha \in [0, 2)$. If we allow the distributions to have a Gaussian part, then we would have the class $J_{\alpha,p}$ defined in [51]. This, in turn, contains important subclasses including the Thorin class (when $p = 1$ and $\alpha = 0$), the Goldie-Steutel-Bondesson class (when $p = 1$ and $\alpha = -1$), the class of type M distributions (when $p = 2$ and $\alpha = 0$), and the class of type G distributions (when $p = 2$ and $\alpha = -1$). For more information about these classes see the references in [6] and [3].

An important feature of p-tempered α-stable distributions is that their associated Lévy processes are multiscaling, i.e. their behavior in a short time frame may be very different from their behavior in a long time frame. This is in contrast to α-stable Lévy processes, which have the same type of behavior at all time scales.

Multiscaling behavior is observed in many applications. For example, evidence of multiscaling behavior in human mobility is given in [26, 61, 64], and [16], while evidence of multiscaling behavior in animal foraging patterns is given in [43] and the references therein. Another important example comes from financial modeling, where it has been observed that very frequent returns (at say the half-hour level) are often well approximated by infinite variance stable distributions. However, returns tend to exhibit aggregational Gaussianity and at large aggregation levels (say at the weekly level) they are well approximated by the Gaussian, see [21] or [35].

The main topic of this brief is the study of p-tempered α-stable distributions and the multiscaling behavior of their associated Lévy processes. Toward this end we begin in Chapter 2 by reviewing some background topics. In Chapter 3 we formally introduce p-tempered α-stable distributions and discuss many properties. It turns out that the class TS^p_α is not closed under weak convergence. In Chapter 4 we introduce the closure of this class and characterize weak convergence in it. In Chapter 5 we characterize the multiscale properties of p-tempered α-stable Lévy processes. Then in Chapters 6 and 7 we give some examples and applications. In particular, in Chapter 6 we explore a number of parametric classes of p-tempered α-stable distributions, and in Chapter 7 we discuss applications to mathematical finance and to mobility models. Further, we discuss a theoretical mechanism by which p-tempered α-stable distributions appear in applications.

Chapter 2
Preliminaries

In this chapter we bring together background material on several topics that will be important in the sequel.

2.1 Basic Topology

In this section we review some basic concepts from topology. For more details see, e.g., [13] or Chapter 7 in [8]. We begin by defining a topological space.

Definition 2.1. Let $\mathbb{E}$ be a set. If $\mathscr{T}$ is a collection of subsets of $\mathbb{E}$ such that

1. $\emptyset, \mathbb{E} \in \mathscr{T}$,
2. $\mathscr{T}$ is closed under finite intersections, and
3. $\mathscr{T}$ is closed under arbitrary unions,

then $\mathscr{T}$ is called a **topology**, $(\mathbb{E}, \mathscr{T})$ is called a **topological space**, and the sets in $\mathscr{T}$ are called **open sets**. The complement of an open set is called a **closed set**. If, in addition, for any $a, b \in \mathbb{E}$ with $a \neq b$ there are $A, B \in \mathscr{T}$ with $a \in A$, $b \in B$, and $A \cap B = \emptyset$, then we say that the space is **Hausdorff**.

For any topological space $(\mathbb{E}, \mathscr{T})$ the class of **Borel sets** is the σ-algebra generated by $\mathscr{T}$ and is denoted by $\mathfrak{B}(\mathbb{E}, \mathscr{T})$. Thus $\mathfrak{B}(\mathbb{E}, \mathscr{T}) = \sigma(\mathscr{T})$. When the collection $\mathscr{T}$ is clear from context we sometimes write $\mathfrak{B}(\mathbb{E}) = \mathfrak{B}(\mathbb{E}, \mathscr{T})$. In particular, when working with $\mathbb{R}^d$ we generally assume that $\mathscr{T}$ are the usual open sets. In this case $\mathfrak{B}(\mathbb{R}^d, \mathscr{T})$ are the usual Borel sets, which we denote by $\mathfrak{B}(\mathbb{R}^d)$. Any measure on the space $(\mathbb{E}, \mathfrak{B}(\mathbb{E}))$ is called a Borel measure on $(\mathbb{E}, \mathscr{T})$ or just a Borel measure when the space is clear from context.

If $A \subset \mathbb{E}$, then the **interior** of A (denoted A°) is the union of all open sets contained in A, and the **closure** of A (denoted $\bar{A}$) is the intersection of all closed

M. Grabchak, *Tempered Stable Distributions*, SpringerBriefs in Mathematics, DOI 10.1007/978-3-319-24927-8_2

sets containing A. Note that $A^\circ \subset A \subset \bar{A}$. We write $\partial A = \bar{A} \setminus A^\circ$ to denote the **boundary** of A. We conclude this section by recalling the definition of a compact set.

Definition 2.2. Let $(\mathbb{E}, \mathscr{T})$ be a Hausdorff space and let $A \subset \mathbb{E}$. If for any collection $\mathscr{T}_0 \subset \mathscr{T}$ with $A \subset \bigcup \mathscr{T}_0$ there is a finite subcollection $\mathscr{T}_1 \subset \mathscr{T}_0$ with $A \subset \bigcup \mathscr{T}_1$, then A is called a **compact set**. If A is such that its closure is compact, then A is called **relatively compact**.

2.2 Infinitely Divisible Distributions and Lévy Processes

In this section we review some important results about infinitely divisible distributions and their associated Lévy processes. Comprehensive references are [69] and [21]. A probability measure μ on $\mathbb{R}^d$ is called **infinitely divisible** if for any positive integer n there exists a probability measure μ_n on $\mathbb{R}^d$ such that if $X \sim \mu$ and $Y_1^{(n)}, \ldots, Y_n^{(n)} \overset{\text{iid}}{\sim} \mu_n$ then

$$X \overset{d}{=} \sum_{i=1}^{n} Y_i^{(n)}.$$

We denote the class of infinitely divisible distributions by ID. The characteristic function of an infinitely divisible distribution μ on $\mathbb{R}^d$ is given by $\hat{\mu}(z) = \exp\{C_\mu(z)\}$ where

$$C_\mu(z) = -\frac{1}{2}\langle z, Az\rangle + i\langle b, z\rangle + \int_{\mathbb{R}^d} \left(e^{i\langle z,x\rangle} - 1 - i\frac{\langle z, x\rangle}{1+|x|^2} \right) M(\mathrm{d}x), \quad (2.1)$$

A is a symmetric nonnegative-definite $d \times d$ matrix, $b \in \mathbb{R}^d$, and M satisfies

$$M(\{0\}) = 0 \text{ and } \int_{\mathbb{R}^d} (|x|^2 \wedge 1) M(\mathrm{d}x) < \infty. \quad (2.2)$$

We call C_μ the **cumulant generating function** of μ, A the **Gaussian part**, b the **shift**, and M the **Lévy measure**. The measure μ is uniquely identified by the **Lévy triplet** (A, M, b) and we will write

$$\mu = ID(A, M, b).$$

The class of infinitely divisible distributions is intimately related with the class of Lévy processes. These processes are defined as follows.

Definition 2.3. A stochastic process $\{X_t : t \geq 0\}$ on $(\Omega, \mathscr{F}, P)$ with values in $\mathbb{R}^d$ is called a **Lévy Process** if $X_0 = 0$ a.s. and the following conditions are satisfied:

1. (Independent increments) For any $n \geq 1$ and $0 \leq t_0 < t_1 < \cdots < t_n < \infty$, the random variables $X_{t_0}, X_{t_1} - X_{t_0}, \ldots, X_{t_n} - X_{t_{n-1}}$ are independent.
2. (Stationary increments) $X_{s+t} - X_s \stackrel{d}{=} X_t$ for any $s, t \geq 0$.
3. (Stochastic continuity) For every $t \geq 0$ and $\epsilon > 0$ $\lim_{s \to t} P(|X_s - X_t| > \epsilon) = 0$.
4. (Càdlàg paths) There is $\Omega_0 \in \mathcal{F}$ with $P(\Omega_0) = 1$ such that for every $\omega \in \Omega_0$, $X_t(\omega)$ is right-continuous in $t \geq 0$ and has left limits in $t > 0$.

Since a Lévy process $\{X_t : t \geq 0\}$ has the càdlàg paths property it follows that, with probability 1, $\lim_{s \downarrow t} X_s = X_t$ and $\lim_{s \uparrow t} X_s$ exists. We define $X_{t-} := \lim_{s \uparrow t} X_s$ and we write $\Delta X_t = X_t - X_{t-}$ to denote the jump at time t. The connection between Lévy processes and infinitely divisible distributions is highlighted by the following result, which is given in Theorem 7.10 of [69].

Proposition 2.4. *1. If μ is an infinitely divisible distribution on $\mathbb{R}^d$, then there is a Lévy process $\{X_t : t \geq 0\}$ with $X_1 \sim \mu$.*
2. Conversely, if $\{X_t : t \geq 0\}$ is a Lévy process on $\mathbb{R}^d$, then for any $t \geq 0$ the distribution μ_t of X_t is infinitely divisible and $\hat{\mu}_t(z) = [\hat{\mu}_1(z)]^t$.
3. If $\{X_t : t \geq 0\}$ and $\{X'_t : t \geq 0\}$ are Lévy processes on $\mathbb{R}^d$ with $X_1 \stackrel{d}{=} X'_1$, then $\{X_t : t \geq 0\}$ and $\{X'_t : t \geq 0\}$ have the same finite dimensional distributions.

In the context of Lévy processes, the Lévy measure has a simple interpretation. Specifically, if $\{X_t : t \geq 0\}$ is a Lévy process with $X_1 \sim ID(A, M, b)$, then

$$M(B) = \mathrm{E}\left[\#\{t \in [0,1] : \Delta X_t \neq 0, \Delta X_t \in B\}\right], \quad B \in \mathfrak{B}(\mathbb{R}^d). \tag{2.3}$$

In other words, $M(B)$ is the expected number of times $t \in [0,1]$ at which the Lévy process has a jump (i.e., $X_t - X_{t-} \neq 0$) and the value of this jump is in the set B. See Sections 3.3–3.4 in [21] for details.

An important subclass of infinitely divisible distributions is the class of stable distributions. A probability measure μ on $\mathbb{R}^d$ is called **stable** if for any n and any $X_1, \ldots, X_n \stackrel{\text{iid}}{\sim} \mu$ there are $a_n > 0$ and $b_n \in \mathbb{R}^d$ such that

$$X_1 \stackrel{d}{=} a_n \sum_{k=1}^{n} X_k - b_n. \tag{2.4}$$

It turns out that, necessarily, $a_n = n^{-1/\alpha}$ for some $\alpha \in (0,2]$. We call this parameter the **index of stability** and we refer to any stable distribution with index α as **α-stable**. Comprehensive references are [68] and [78].

Fix $\alpha \in (0,2]$ and let μ be an α-stable distribution. If $\alpha = 2$, then $\mu = ID(A, 0, b)$ is a multivariate normal distribution, which we denote by $\mu = N(b, A)$. If $\alpha \in (0,2)$, then $\mu = ID(0, L, b)$ where

$$L(A) = \int_{\mathbb{S}^{d-1}} \int_0^\infty 1_A(ur) r^{-1-\alpha} dr \sigma(du), \qquad A \in \mathfrak{B}(\mathbb{R}^d),$$

for some finite Borel measure σ on $\mathbb{S}^{d-1}$. We call σ the **spectral measure** of the distribution and we write $\mu = S_\alpha(\sigma, b)$. All α-stable distributions with $\alpha \in (0, 2)$ and $\sigma \neq 0$ have an infinite variance and are sometimes called **infinite variance stable distributions**.

One reason for the importance of stable distributions is that they are the only possible limits of scaled and shifted sums of iid random variables. Specifically, let $X_1, X_2, \cdots \stackrel{\text{iid}}{\sim} \mu$ for some probability measure μ and define $S_n = \sum_{i=1}^n X_i$. If there exists a probability measure ν and sequences $a_n > 0$ and $b_n \in \mathbb{R}^d$ such that for $Y \sim \nu$

$$(a_n S_n - b_n) \stackrel{d}{\to} Y, \tag{2.5}$$

then ν is a stable distribution. When this holds we say that μ (or equivalently X_1) belongs to **the domain of attraction** of ν (or equivalently of Y). When ν is not degenerate its domain of attraction is characterized in [23] for the case $d = 1$ and in [67] and [54] for the case $d \geq 2$. We now give a related fact, which further explains the importance of stable distributions.

Lemma 2.5. *Fix $c \in \{0, \infty\}$. Let $\{X_t : t \geq 0\}$ be a Lévy process and let Y be a random variable whose distribution is not concentrated at a point. If there exist functions $a_t > 0$ and $b_t \in \mathbb{R}^d$ with*

$$(a_t X_t - b_t) \stackrel{d}{\to} Y \text{ as } t \to c \tag{2.6}$$

then Y has an α-stable distribution for some $\alpha \in (0, 2]$.

Proof. Fix $N \in \mathbb{N}$. Let $Y^{(1)}, Y^{(2)}, \ldots, Y^{(N)}$ be iid copies of Y and let $\{X_t^{(n)} : t \geq 0\}$, $n = 1, 2, \ldots, N$, be independent Lévy processes with $X_1^{(n)} \stackrel{d}{=} X_1$. From (2.6) it follows that

$$\operatorname*{d\text{-}lim}_{t \to c} (a_{Nt} X_{Nt} - b_{Nt}) = Y.$$

The fact that Lévy processes have independent and stationary increments gives

$$\begin{aligned}\operatorname*{d\text{-}lim}_{t \to c} (a_t X_{Nt} - N b_t) &= \operatorname*{d\text{-}lim}_{t \to c} \sum_{n=1}^N \left[a_t \left(X_{nt} - X_{(n-1)t}\right) - b_t\right] \\ &= \operatorname*{d\text{-}lim}_{t \to c} \sum_{n=1}^N \left(a_t X_t^{(n)} - b_t\right) = \sum_{n=1}^N Y^{(n)}.\end{aligned}$$

Since Y is not concentrated at a point, neither is $\sum_{n=1}^N Y^{(n)}$, and by the Convergence of Types Theorem (see, e.g., Lemma 13.10 in [69]) there are constants $c_N > 0$ and $d_N \in \mathbb{R}^d$ such that

$$\sum_{n=1}^N Y^{(n)} \stackrel{d}{=} c_N Y - d_N,$$

which implies that Y has a stable distribution by (2.4). □

2.3 Regular Variation

Regularly varying functions are functions that have power-like behavior. Comprehensive references are [11, 23, 62], and [63]. For $c \in \{0, \infty\}$ and $\rho \in \mathbb{R}$, a Borel function $f : (0, \infty) \mapsto (0, \infty)$ is called **regularly varying at c with index ρ** if

$$\lim_{x \to c} \frac{f(tx)}{f(x)} = t^{\rho}.$$

In this case we write $f \in RV_{\rho}^{c}$. If $f \in RV_{\rho}^{c}$, then there is an $L \in RV_{0}^{c}$ such that $f(x) = x^{\rho} L(x)$. If $h(x) = f(1/x)$, then

$$f \in RV_{\rho}^{c} \text{ if and only if } h \in RV_{-\rho}^{1/c}. \tag{2.7}$$

If $f \in RV_{\rho}^{c}$ with $\rho > 0$ and $f^{\leftarrow}(x) = \inf\{y > 0 : f(y) > x\}$, then

$$f^{\leftarrow} \in RV_{1/\rho}^{c} \tag{2.8}$$

and $f^{\leftarrow}$ is an asymptotic inverse of f in the sense that

$$f(f^{\leftarrow}(x)) \sim f^{\leftarrow}(f(x)) \sim x \text{ as } x \to c.$$

When $c = \infty$ this result is given on page 28 of [11]. The case when $c = 0$ can be shown using an extension of those results and (2.7). We now summarize several important properties of regularly varying functions.

Proposition 2.6. *Fix $c \in \{0, \infty\}$ and $\rho \in \mathbb{R}$. Let $f, g, h : (0, \infty) \mapsto (0, \infty)$.*

1. *If $f \in RV_{\rho}^{c}$, then*

$$\lim_{t \to c} f(t) = \begin{cases} 1/c & \text{if } \rho < 0 \\ c & \text{if } \rho > 0 \end{cases}.$$

2. *If f is a monotone function and there are sequences of positive numbers λ_n and b_n such that $b_n \to c$, $\lim_{n\to\infty} \lambda_n/\lambda_{n+1} = 1$, and if for all $x > 0$*

$$\lim_{n \to \infty} \lambda_n f(b_n x) =: \chi(x) \tag{2.9}$$

exists and is positive and finite, then there is a $\rho \in \mathbb{R}$ such that $\chi(x)/\chi(1) = x^{\rho}$ and $f \in RV_{\rho}^{c}$.
3. *Let $f \in RV_{\rho}^{c}$ and assume that $h(x) \to c$ as $x \to c$. If for some $k > 0$ we have $g(x) \sim kh(x)$ as $x \to c$, then $f(g(x)) \sim k^{\rho} f(h(x))$ as $x \to c$.*
4. *If $k > 0$, $\rho > 0$, and $f, g \in RV_{\rho}^{c}$, then*

$$f(t) \sim kg(t) \text{ as } t \to c$$

if and only if

$$f^{\leftarrow}(t) \sim k^{-1/\rho} g^{\leftarrow}(t) \text{ as } t \to c.$$

Proof. For the case $c = \infty$ Parts 1–3 are given in Propositions 2.3 and 2.6 in [63]. Extensions to the case $c = 0$ follow from (2.7). Part 4 is an immediate consequence of Part 3 and the asymptotic uniqueness of asymptotic inverses of regularly varying functions, see Theorem 1.5.12 in [11]. □

Another useful result is Karamata's Theorem, a version of which is as follows.

Theorem 2.7. *Fix $c \in \{0, \infty\}$ and let $f \in RV_\rho^c$ for some $\rho \in \mathbb{R}$. If $\rho \geq -1$ and $\int_0^x f(t)\mathrm{d}t < \infty$ for all $x > 0$, then*

$$\lim_{x \to c} \frac{xf(x)}{\int_0^x f(t)\mathrm{d}t} = \rho + 1. \tag{2.10}$$

If $\rho \leq -1$ and $\int_x^\infty f(t)\mathrm{d}t < \infty$ for all $x > 0$, then

$$\lim_{x \to c} \frac{xf(x)}{\int_x^\infty f(t)\mathrm{d}t} = -\rho - 1. \tag{2.11}$$

Proof. For $c = \infty$ this follows from Theorem 2.1 in [63]. Now assume that $c = 0$. To verify (2.10) let $g(x) = x^{-2} f(1/x)$ and note that (2.7) implies that $g \in RV_{-2-\rho}^\infty$. By change of variables we have

$$\lim_{x \to 0} \frac{xf(x)}{\int_0^x f(t)\mathrm{d}t} = \lim_{x \to \infty} \frac{x^{-1} f(1/x)}{\int_0^{1/x} f(t)\mathrm{d}t} = \lim_{x \to \infty} \frac{xg(x)}{\int_x^\infty g(t)\mathrm{d}t} = \lim_{x \to \infty} \frac{xg(x)}{\int_x^\infty g(t)\mathrm{d}t} = \rho + 1,$$

where the final equality follows by (2.11) for the case $c = \infty$ and the fact that $-2 - \rho \leq -1$. The proof of (2.11) is similar. □

We will also work with matrix-valued functions. While regular variation of invertible matrix-valued functions is defined in [5] and [54], we need a different definition to allow for the non-invertible case.

Definition 2.8. Fix $c \in \{0, \infty\}$, $\rho \in \mathbb{R}$, and let $A_\bullet : (0, \infty) \mapsto \mathbb{R}^{d \times d}$. If $\mathrm{tr} A_\bullet \in RV_\rho^c$ and there exists a $B \in \mathbb{R}^{d \times d}$ with $B \neq 0$ and

$$\lim_{t \to c} \frac{A_t}{\mathrm{tr} A_t} = B$$

we say that $A_\bullet$ is **matrix regularly varying at c with index ρ and limiting matrix B**. In this case we write $A_\bullet \in MRV_\rho^c(B)$.

In the above definition, we can allow scaling by a function other than $\mathrm{tr} A_\bullet$. However, this choice is convenient for our purposes. One way to interpret matrix regular variation is in terms of quadratic forms. It is straightforward to show that $A_\bullet \in MRV_\rho^c(B)$ means that there exists an $L \in RV_0^c$ such that for any $z \in \mathbb{R}^d$

$$\langle z, A_t z\rangle \sim \langle z, Bz\rangle t^\rho L(t) \text{ as } t \to c. \tag{2.12}$$

We also need to define regular variation for measures. Assume that R is a Borel measure on $\mathbb{R}^d$ with

$$R(|x| > \delta) < \infty \text{ for any } \delta > 0. \tag{2.13}$$

Note that this condition holds for all probability measures and all Lévy measures.

Definition 2.9. Fix $\rho \le 0$ and $c \in \{0, \infty\}$. A Borel measure R on $\mathbb{R}^d$ satisfying (2.13) is said to be **regularly varying at** c **with index** ρ if there exists a finite, non-zero Borel measure σ on $\mathbb{S}^{d-1}$ such that for all $D \in \mathfrak{B}(\mathbb{S}^{d-1})$ with $\sigma(\partial D) = 0$

$$\lim_{r\to c} \frac{R\left(|x| > rt, \frac{x}{|x|} \in D\right)}{R(|x| > r)} = t^\rho \frac{\sigma(D)}{\sigma(\mathbb{S}^{d-1})}. \tag{2.14}$$

When this holds we write $R \in RV_\rho^c(\sigma)$ and we refer to σ as a **limiting measure**.

Clearly, the measure σ is unique only up to a multiplicative constant. For $D \in \mathfrak{B}(\mathbb{S}^{d-1})$ define

$$U_D(t) = R(|x| > t, x/|x| \in D), \qquad t > 0. \tag{2.15}$$

When $\sigma(D) > 0$, $\sigma(\partial D) = 0$, and $R \in RV_\rho^c(\sigma)$

$$\lim_{r\to c} \frac{U_D(rt)}{U_D(r)} = \lim_{r\to c} \frac{U_D(rt)}{U_{\mathbb{S}^{d-1}}(r)} \frac{U_{\mathbb{S}^{d-1}}(r)}{U_D(r)} = t^\rho \frac{\sigma(D)}{\sigma(\mathbb{S}^{d-1})} \frac{\sigma(\mathbb{S}^{d-1})}{\sigma(D)} = t^\rho,$$

and hence

$$U_D \in RV_\rho^c. \tag{2.16}$$

In particular, we have $U_{\mathbb{S}^{d-1}} \in RV_\rho^c$. Now take $L(t) = U_{\mathbb{S}^{d-1}}(t) / \left[t^\rho \sigma(\mathbb{S}^{d-1})\right]$, and note that $L \in RV_0^c$. Combining this with (2.14) gives the following.

Lemma 2.10. *$R \in RV_\rho^c(\sigma)$ if and only if there is an $L \in RV_0^c$ such that for all $D \in \mathfrak{B}(\mathbb{S}^{d-1})$ with $\sigma(\partial D) = 0$*

$$U_D(t) \sim \sigma(D) t^\rho L(t) \text{ as } t \to c. \tag{2.17}$$

The next result will be fundamental to the discussion in Chapter 5.

Proposition 2.11. *Fix $c \in \{0, \infty\}$, $\rho \le 0$, let $\sigma \ne 0$ be a finite Borel measure on $\mathbb{S}^{d-1}$, and let R be a Borel measure on $\mathbb{R}^d$ satisfying* (2.13).

1. *If $R \in RV^c_\rho(\sigma)$ and $q \ge 0$ with $0 < q + |\rho|$, then for any $\kappa > 0$ there exists a function $a_t > 0$ with $\lim_{t\to c} a_t = 1/c$ such that*

$$\lim_{t\to c} t a_t^q R\left(|x| > r/a_t, \frac{x}{|x|} \in D\right) = \kappa\sigma(D) r^\rho \tag{2.18}$$

for all $r \in (0,\infty)$ and all $D \in \mathfrak{B}(\mathbb{S}^{d-1})$ with $\sigma(\partial D) = 0$.

2. *If there exists a function $a_t > 0$ with $\lim_{t\to c} a_t = 1/c$ such that for all $r \in (0,\infty)$ and all $D \in \mathfrak{B}(\mathbb{S}^{d-1})$ with $\sigma(\partial D) = 0$* (2.18) *holds for some $q \ge 0$ and some $\kappa > 0$, then $R \in RV^c_\rho(\sigma)$.*
3. *If $R \in RV^c_\rho(\sigma)$ and $q \ge 0$ with $0 < q + |\rho|$, then* (2.18) *holds for some function $a_t > 0$ with $\lim_{t\to c} a_t = 1/c$ if and only if $a_t \sim K^{1/(|\rho|+q)}/V^{\leftarrow}(t)$ where $K = \kappa\sigma(\mathbb{S}^{d-1})$ and $V(t) = t^q/R(|x| > t)$. Moreover, in this case, $a_\bullet \in RV^c_{-1/(q+|\rho|)}$.*

Proof. Fix $D \in \mathfrak{B}(\mathbb{S}^{d-1})$ with $\sigma(\partial D) = 0$. We begin with the first part. Assume that $R \in RV^c_\rho(\sigma)$ and let $a_t \sim K^{1/(|\rho|+q)}/V^{\leftarrow}(t)$, where V and K are as in Part 3. Note that $a_\bullet \in RV^c_{-1/(q+|\rho|)}$ and thus that $\lim_{t\to c} a_t = 1/c$. By Proposition 2.6 we have

$$\begin{aligned}
r^\rho \frac{\sigma(D)}{\sigma(\mathbb{S}^{d-1})} &= \lim_{s\to c} \frac{R\left(|x| > rs, \frac{x}{|x|} \in D\right)}{R(|x| > s)} \\
&= \lim_{t\to c} \frac{a_t^q R\left(|x| > r/a_t, \frac{x}{|x|} \in D\right)}{a_t^q R(|x| > 1/a_t)} \\
&= \lim_{t\to c} V(1/a_t) a_t^q R\left(|x| > r/a_t, \frac{x}{|x|} \in D\right) \\
&= K^{-1} \lim_{t\to c} V(K^{1/(|\rho|+q)}/a_t) a_t^q R\left(|x| > r/a_t, \frac{x}{|x|} \in D\right) \\
&= K^{-1} \lim_{t\to c} V(V^{\leftarrow}(t)) a_t^q R\left(|x| > r/a_t, \frac{x}{|x|} \in D\right) \\
&= \frac{1}{\kappa\sigma(\mathbb{S}^{d-1})} \lim_{t\to c} t a_t^q R\left(|x| > r/a_t, \frac{x}{|x|} \in D\right)
\end{aligned}$$

as required. To show the second part assume that (2.18) holds for some $q \ge 0$, some $\kappa > 0$, and some function $a_t > 0$ satisfying $\lim_{t\to c} a_t = 1/c$. We have

$$\lim_{s\to c} \frac{R\left(|x| > sr, \frac{x}{|x|} \in D\right)}{R(|x| > s)} = \lim_{t\to c} \frac{t a_t^q R\left(|x| > r/a_t, \frac{x}{|x|} \in D\right)}{t a_t^q R(|x| > 1/a_t)} = \frac{\sigma(D)}{\sigma(\mathbb{S}^{d-1})} r^\rho.$$

We now turn to the third part. Assume that $a_t > 0$ is such that $\lim_{t\to c} a_t = 1/c$ and that $a_\bullet$ satisfies (2.18) for all $r \in (0, \infty)$ and all $D \in \mathfrak{B}(\mathbb{S}^{d-1})$ with $\sigma(\partial D) = 0$. In particular, this means that $\lim_{t\to c} ta_t^q R(|x| > 1/a_t) = \kappa\sigma(\mathbb{S}^{d-1})$, or equivalently that $V(1/a_t) \sim t/K$ as $t \to c$. Combining this with Proposition 2.6 gives

$$\lim_{t\to c} \frac{a_t}{K^{1/(|\rho|+q)}/V^{\leftarrow}(t)} = \lim_{t\to c} \frac{K^{-1/(|\rho|+q)}V^{\leftarrow}(t)}{1/a_t}$$

$$= \lim_{t\to c} \frac{V^{\leftarrow}(t/K)}{V^{\leftarrow}(V(1/a_t))} = \lim_{t\to c} \frac{V^{\leftarrow}(t/K)}{V^{\leftarrow}(t/K)} = 1,$$

which concludes the proof. □

When $R \in RV_\rho^\infty(\sigma)$ we sometimes say that R has **regularly varying tails**. In this case we refer to $|\rho|$ as the **tail index**. The following result helps explain these definitions.

Proposition 2.12. *Let $\sigma \neq 0$ be a finite Borel measure on $\mathbb{S}^{d-1}$ and let R be a Borel measure on $\mathbb{R}^d$ satisfying (2.13). If $R \in RV_\rho^\infty(\sigma)$ for some $\rho \le 0$, then for any $\delta > 0$*

$$\int_{|x|\ge\delta} |x|^\gamma R(\mathrm{d}x) \begin{cases} < \infty & \text{if } \gamma < |\rho| \\ = \infty & \text{if } \gamma > |\rho| \end{cases}.$$

Proof. When $\gamma \le 0$ the result follows immediately from the fact that R satisfies (2.13). Now assume that $\gamma > 0$ and fix $\delta > 0$. By Fubini's Theorem (Theorem 18.3 in [10])

$$\int_{|x|\ge\delta} |x|^\gamma R(\mathrm{d}x) = \int_{|x|\ge\delta}\int_0^{|x|} \gamma u^{\gamma-1}\mathrm{d}uR(\mathrm{d}x)$$

$$= \delta^\gamma R(|x| \ge \delta) + \int_\delta^\infty \gamma u^{\gamma-1} R(|x| \ge u)\mathrm{d}u = I_1 + I_2.$$

Clearly, $I_1 < \infty$. From (2.16) it follows that $R(|x| \ge u) = u^\rho L(u)$ for some $L \in RV_0^\infty$. Proposition 1.3.6 in [11] implies that for any $\epsilon > 0$ there exists a $\delta_\epsilon > \delta$ such that for all $u > \delta_\epsilon$ we have $u^{-\epsilon} < L(u) < u^\epsilon$. When $\gamma > |\rho|$ fix $\epsilon \in (0, \gamma - |\rho|)$ and note that

$$I_2 \ge \int_{\delta_\epsilon}^\infty \gamma u^{\gamma-1-|\rho|-\epsilon}\mathrm{d}u = \infty.$$

When $\gamma < |\rho|$ fix $\epsilon \in (0, |\rho| - \gamma)$ and note that

$$I_2 \le \int_\delta^{\delta_\epsilon} \gamma u^{\gamma-1} R(|x| \ge u)\mathrm{d}u + \int_{\delta_\epsilon}^\infty \gamma u^{\gamma-|\rho|+\epsilon-1}\mathrm{d}u < \infty.$$

This completes the proof. □

Chapter 3
Tempered Stable Distributions

In this chapter we formally define tempered stable distributions and discuss many properties. These distributions were first introduced in [65]. From here the class was expanded in several directions in [9, 51, 66], and [27]. Our discussion mainly follows [27].

3.1 Definitions and Basic Properties

Fix $\alpha \in (0, 2)$, let σ be a finite Borel measure on $\mathbb{S}^{d-1}$, and recall that the Lévy measure of an α-stable distribution with spectral measure σ is given by

$$L(A) = \int_{\mathbb{S}^{d-1}} \int_0^\infty 1_A(ru) r^{-\alpha-1} \mathrm{d}r\sigma(\mathrm{d}u), \quad A \in \mathfrak{B}(\mathbb{R}^d). \tag{3.1}$$

Now, fix $p > 0$ and define a new Lévy measure of the form

$$M(A) = \int_{\mathbb{S}^{d-1}} \int_0^\infty 1_A(ru) q(r^p, u) r^{-\alpha-1} \mathrm{d}r\sigma(\mathrm{d}u), \quad A \in \mathfrak{B}(\mathbb{R}^d), \tag{3.2}$$

where $q : (0, \infty) \times \mathbb{S}^{d-1} \mapsto (0, \infty)$ is a Borel function such that, for all $u \in \mathbb{S}^{d-1}$, $q(\cdot, u)$ is completely monotone and satisfies

$$\int_0^1 r^{1-\alpha} q(r^p, u)\mathrm{d}r < \infty, \quad \int_1^\infty r^{-1-\alpha} q(r^p, u)\mathrm{d}r < \infty, \tag{3.3}$$

and

$$\lim_{r\to\infty} q(r, u) = 0. \tag{3.4}$$

M. Grabchak, *Tempered Stable Distributions*, SpringerBriefs in Mathematics, DOI 10.1007/978-3-319-24927-8_3

The conditions in (3.3) guarantee that this is a valid Lévy measure, while the fact that (3.4) holds implies that the tails of M are lighter than those of L. This implies that the tails of the associated infinitely divisible distribution are lighter as well.

The complete monotonicity[1] of $q(\cdot, u)$ means that, for each $u \in \mathbb{S}^{d-1}$, the function $q(r, u)$ is infinitely differentiable in r and

$$(-1)^n \frac{\partial^n}{\partial r^n} q(r, u) \geq 0. \tag{3.5}$$

In particular, this implies that $q(\cdot, u)$ is a monotonely decreasing function for each $u \in \mathbb{S}^{d-1}$. By (3.4) and Bernstein's Theorem (see, e.g., Theorem 1a in Section XIII.4 of [23] or Remark 3.2 in [6]) it follows that there exists a measurable family[2] $\{Q_u\}_{u \in \mathbb{S}^{d-1}}$ of Borel measures on $(0, \infty)$ with

$$q(r^p, u) = \int_{(0,\infty)} e^{-r^p s} Q_u(\mathrm{d}s). \tag{3.6}$$

From here it follows that, so long as $Q_u \neq 0$, we have $q(r^p, u) > 0$ for all $r > 0$.

Note that, under the given conditions on the function q, (3.2) defines a valid Lévy measure even for α outside of the interval $(0, 2)$. However, since $q(\cdot, u)$ is a decreasing function for each $u \in \mathbb{S}^{d-1}$, when $\alpha \geq 2$ condition (3.3) holds only with $q \equiv 0$. For this reason, we only consider the case $\alpha \in (-\infty, 2)$. This leads to the following definition.

Definition 3.1. Fix $\alpha < 2$ and $p > 0$. An infinitely divisible probability measure μ is called a **p-tempered α-stable distribution** if it has no Gaussian part and its Lévy measure is given by (3.2), where σ is a finite Borel measure on $\mathbb{S}^{d-1}$ and $q : (0, \infty) \times \mathbb{S}^{d-1} \mapsto (0, \infty)$ is a Borel function such that for all $u \in \mathbb{S}^{d-1}$, $q(\cdot, u)$ is completely monotone and satisfies (3.3) and (3.4). We denote the class of p-tempered α-stable distributions by TS^p_α.

We use the term **tempered stable distributions** to refer to the class of all p-tempered α-stable distributions with all $\alpha < 2$ and $p > 0$.

Remark 3.1. Under appropriate integrability conditions, one can define Lévy measures of the form (3.2) with $p \leq 0$. The case $p = 0$ corresponds to the class of α-stable distributions and only makes sense for $\alpha \in (0, 2)$. The case $p < 0$ has significantly different behavior from the case $p > 0$ and will not be considered here.

Remark 3.2. From Theorem 15.10 in [69] it follows that p-tempered α-stable distributions belong to the class of self-decomposable distributions if and only if $q(r^p, u)r^{-\alpha}$ is a decreasing function of r for every $u \in \mathbb{S}^{d-1}$. This always holds when

---

[1] A general reference on completely monotone functions is [72].

[2] The measurability of the family means that for any Borel set A the function $f(u) = Q_u(A)$ is measurable.

$\alpha \in [0, 2)$, but it may fail when $\alpha < 0$. Thus, when $\alpha \in [0, 2)$, p-tempered α-stable distributions possess all properties of self-decomposable distributions. In particular, if they are nondegenerate, then they have a density with respect to Lebesgue measure in d-dimensions and when $d = 1$ they are unimodal.

In Definition 3.1, the case when $\alpha \leq 0$ no longer corresponds to the idea of modifying the tails of a stable distribution. Nevertheless, such distributions serve to make the class richer and more robust. It should be added that, even in the case when $\alpha \in (0, 2)$ we may no longer have a Lévy measure that looks much like that of an α-stable distribution. For that to hold, we would need the function q to be close to 1 in some region near zero. This leads to the following definition.

Definition 3.2. Fix $p > 0$ and $\alpha < 2$. Let μ be a p-tempered α-stable distribution with Lévy measure M. If M can be represented in the form (3.2) where

$$\lim_{r \downarrow 0} q(r, u) = 1 \text{ for every } u \in \mathbb{S}^{d-1}, \tag{3.7}$$

then μ is called a **proper p-tempered α-stable distribution**.

Proper p-tempered α-stable distributions with $\alpha \in (0, 2)$ are the ones that correspond to the original motivation of modifying the tails of stable distributions to make them lighter.

Remark 3.3. In [9, 65], and [27] proper p-tempered α-stable distributions are defined to be ones where M is of the form (3.2) and (3.7) holds. However, it may happen that q does not satisfy (3.7), but that there is a Borel function $c : \mathbb{S}^{d-1} \mapsto (0, \infty)$ such that $q'(r, u) = q(r, u)/c(u)$ satisfies (3.7). In this case we can take $\sigma'(du) = c(u)\sigma(du)$ and write M as (3.2) but with q' and σ' in place of q and σ. In this case we still want to consider M to be a proper p-tempered α-stable distribution. For this reason we need the somewhat more subtle formulation given in Definition 3.2.

Remark 3.4. Assume that $q(r, u)$ satisfies (3.6). The Monotone Convergence Theorem implies that $q(r, u)$ satisfies (3.7) if and only if (3.6) holds with Q_u being a *probability* measure for every $u \in \mathbb{S}^{d-1}$.

Remark 3.5. When $\alpha \in (0, 2)$ and $p > 0$, the class of proper p-tempered α-stable distributions belongs to the class of Generalized Tempered Stable Distributions introduced in [66].

It is somewhat artificial to work with the family of measures $\{Q_u\}_{u \in \mathbb{S}^{d-1}}$ and the measure σ separately. Ideally, we would like to combine these into one object. Toward this end, let Q be a Borel measure on $\mathbb{R}^d$ given by

$$Q(A) = \int_{\mathbb{S}^{d-1}} \int_{(0,\infty)} 1_A(ru) Q_u(dr) \sigma(du), \qquad A \in \mathfrak{B}(\mathbb{R}^d), \tag{3.8}$$

and note that $Q(\{0\}) = 0$. Now define a Borel measure R on $\mathbb{R}^d$ by

$$R(A) = \int_{\mathbb{R}^d} 1_A\left(\frac{x}{|x|^{1+1/p}}\right)|x|^{\alpha/p}Q(\mathrm{d}x), \qquad A \in \mathfrak{B}(\mathbb{R}^d), \tag{3.9}$$

and again note that $R(\{0\}) = 0$. To get the inverse transformation we have

$$Q(A) = \int_{\mathbb{R}^d} 1_A\left(\frac{x}{|x|^{p+1}}\right)|x|^{\alpha}R(\mathrm{d}x), \qquad A \in \mathfrak{B}(\mathbb{R}^d). \tag{3.10}$$

From here it follows that

$$Q(\mathbb{R}^d) = \int_{\mathbb{R}^d} |x|^{\alpha}R(\mathrm{d}x). \tag{3.11}$$

We now write the Lévy measure M in terms of R. By (3.2) and (3.6) for any $A \in \mathfrak{B}(\mathbb{R}^d)$ we have

$$\begin{aligned} M(A) &= \int_{\mathbb{S}^{d-1}}\int_{(0,\infty)}\int_0^{\infty} 1_A(ru)r^{-\alpha-1}e^{-r^p s}\mathrm{d}rQ_u(\mathrm{d}s)\sigma(\mathrm{d}u) \\ &= \int_{\mathbb{S}^{d-1}}\int_{(0,\infty)}\int_0^{\infty} 1_A(ts^{-1/p}u)t^{-1-\alpha}e^{-t^p}\mathrm{d}ts^{\alpha/p}Q_u(\mathrm{d}s)\sigma(\mathrm{d}u) \\ &= \int_{\mathbb{R}^d}\int_0^{\infty} 1_A\left(t\frac{x}{|x|^{1+1/p}}\right)t^{-1-\alpha}e^{-t^p}\mathrm{d}t|x|^{\alpha/p}Q(\mathrm{d}x), \end{aligned}$$

where the second equality follows by the substitution $t = rs^{1/p}$. From here (3.10) gives

$$M(A) = \int_{\mathbb{R}^d}\int_0^{\infty} 1_A(tx)t^{-1-\alpha}e^{-t^p}\mathrm{d}tR(\mathrm{d}x), \quad A \in \mathfrak{B}(\mathbb{R}^d). \tag{3.12}$$

This is the form of the Lévy measure that tends to be the most convenient to work with.

This representation raises several questions: If we are given a measure of the form (3.12), under what conditions will it be a Lévy measure? When it is a Lévy measure, is it necessarily the Lévy measure of a p-tempered α-stable distribution? Is there a one-to-one relationship between the measures M and R? The answers are provided by the following.

Theorem 3.3. *1. Fix $p > 0$ and let M be given by* (3.12)*. M is the Lévy measure of an infinitely divisible distribution if and only if either $\alpha \in \mathbb{R}$ and $R = 0$ or $\alpha < 2$,*

$$R(\{0\}) = 0, \tag{3.13}$$

and

$$\int_{\mathbb{R}^d} \left(|x|^2 \wedge |x|^\alpha\right) R(\mathrm{d}x) < \infty \text{ if } \alpha \in (0,2),$$

$$\int_{\mathbb{R}^d} \left(|x|^2 \wedge [1 + \log^+ |x|]\right) R(\mathrm{d}x) < \infty \quad \text{if } \alpha = 0, \tag{3.14}$$

$$\int_{\mathbb{R}^d} \left(|x|^2 \wedge 1\right) R(\mathrm{d}x) < \infty \quad \text{if } \alpha < 0.$$

2. *Fix $p > 0$, $\alpha < 2$, and let M be given by* (3.12). *If R satisfies* (3.13) *and* (3.14), *then M is the Lévy measure of a p-tempered α-stable distribution and it uniquely determines R. Moreover, M is the Lévy measure of a proper p-tempered α-stable distribution if and only if*

$$\int_{\mathbb{R}^d} |x|^\alpha R(\mathrm{d}x) < \infty. \tag{3.15}$$

Proof. We begin with Part 1. By (2.2) M is a Lévy measure if and only if $M(\{0\}) = 0$ and $\int_{\mathbb{R}^d}(|x|^2 \wedge 1)M(\mathrm{d}x) < \infty$. Assume $R \neq 0$, since the other case is trivial. For any $\alpha \in \mathbb{R}$

$$M(\{0\}) = \int_{\mathbb{R}^d}\int_0^\infty 1_{\{0\}}(tx)t^{-\alpha-1}e^{-t^p}\mathrm{d}tR(\mathrm{d}x) = \int_{\{0\}}\int_0^\infty t^{-1-\alpha}e^{-t^p}\mathrm{d}tR(\mathrm{d}x),$$

which equals zero if and only if $R(\{0\}) = 0$.

Now assume that $\int_{\mathbb{R}^d}(|x|^2 \wedge 1)M(\mathrm{d}x) < \infty$. We will show that this implies that $\alpha < 2$ and that R satisfies (3.14). Fix $\epsilon > 0$ and note that

$$\begin{aligned}\infty > \int_{|x|\leq 1} |x|^2 M(\mathrm{d}x) &= \int_{\mathbb{R}^d} |x|^2 \int_0^{|x|^{-1}} t^{1-\alpha}e^{-t^p}\mathrm{d}tR(\mathrm{d}x)\\ &\geq \int_{|x|\leq 1/\epsilon} |x|^2 \int_0^\epsilon t^{1-\alpha}e^{-t^p}\mathrm{d}tR(\mathrm{d}x) \geq e^{-\epsilon^p}\int_{|x|\leq 1/\epsilon} |x|^2R(\mathrm{d}x)\int_0^\epsilon t^{1-\alpha}\mathrm{d}t.\end{aligned}$$

Since $R \neq 0$, for this be finite for all $\epsilon > 0$ it is necessary that $\alpha < 2$. Taking $\epsilon = 1$ gives the necessity of $\int_{|x|\leq 1} |x|^2R(\mathrm{d}x) < \infty$. Observing that

$$\begin{aligned}\infty > \int_{|x|\geq 1} M(\mathrm{d}x) &= \int_{\mathbb{R}^d}\int_{|x|^{-1}}^\infty t^{-1-\alpha}e^{-t^p}\mathrm{d}tR(\mathrm{d}x)\\ &\geq \int_1^\infty t^{-1-\alpha}e^{-t^p}\mathrm{d}t\int_{|x|\geq 1} R(\mathrm{d}x) + e^{-1}\int_{|x|\geq 1}\int_{|x|^{-1}}^1 t^{-1-\alpha}\mathrm{d}tR(\mathrm{d}x)\end{aligned}$$

gives the necessity of $\int_{|x|\geq 1} R(\mathrm{d}x) < \infty$ and $\int_{|x|\geq 1}\int_{|x|^{-1}}^{1} t^{-1-\alpha}\mathrm{d}tR(\mathrm{d}x) < \infty$. When $\alpha < 0$ we are done. When $\alpha = 0$ we have

$$\int_{|x|\geq 1}\int_{|x|^{-1}}^{1} t^{-1-\alpha}\mathrm{d}tR(\mathrm{d}x) = \int_{|x|\geq 1} \log|x|R(\mathrm{d}x),$$

and when $\alpha \in (0,2)$ we have

$$\int_{|x|\geq 1}\int_{|x|^{-1}}^{1} t^{-1-\alpha}\mathrm{d}tR(\mathrm{d}x) = \frac{1}{\alpha}\int_{|x|\geq 1} \left(|x|^{\alpha}-1\right)R(\mathrm{d}x),$$

which together with the necessity of $\int_{|x|\geq 1} R(\mathrm{d}x) < \infty$ gives (3.14).

Now assume that $\alpha < 2$ and that R satisfies (3.14). We have

$$\begin{aligned}
\int_{|x|\leq 1} |x|^2 M(\mathrm{d}x) &= \int_{\mathbb{R}^d} |x|^2 \int_0^{|x|^{-1}} t^{1-\alpha}e^{-t^p}\mathrm{d}tR(\mathrm{d}x)\\
&\leq \int_{|x|\leq 1} |x|^2 R(\mathrm{d}x)\int_0^{\infty} t^{1-\alpha}e^{-t^p}\mathrm{d}t + \int_{|x|>1}|x|^2\int_0^{|x|^{-1}} t^{1-\alpha}\mathrm{d}tR(\mathrm{d}x)\\
&= p^{-1}\Gamma\left(\frac{2-\alpha}{p}\right)\int_{|x|\leq 1}|x|^2R(\mathrm{d}x) + (2-\alpha)^{-1}\int_{|x|>1}|x|^{\alpha}R(\mathrm{d}x),
\end{aligned}$$

which is finite. Now let $D = \sup_{t\geq 1} t^{2-\alpha}e^{-t^p}$ and note that

$$\begin{aligned}
\int_{|x|\geq 1} M(\mathrm{d}x) &= \int_{\mathbb{R}^d}\int_{|x|^{-1}}^{\infty} t^{-1-\alpha}e^{-t^p}\mathrm{d}tR(\mathrm{d}x)\\
&\leq D\int_{|x|\leq 1}\int_{|x|^{-1}}^{\infty} t^{-3}\mathrm{d}tR(\mathrm{d}x) + \int_{|x|>1}\int_{|x|^{-1}}^{\infty} t^{-1-\alpha}e^{-t^p}\mathrm{d}tR(\mathrm{d}x)\\
&= .5D\int_{|x|\leq 1}|x|^2R(\mathrm{d}x) + \int_{|x|>1}\int_{|x|^{-1}}^{1} t^{-1-\alpha}e^{-t^p}\mathrm{d}tR(\mathrm{d}x)\\
&\quad + \int_1^{\infty} t^{-1-\alpha}e^{-t^p}\mathrm{d}t\int_{|x|>1}R(\mathrm{d}x),
\end{aligned}$$

which is finite since the second integral is bounded by $\int_{|x|>1}\frac{|x|^{\alpha}-1}{\alpha}R(\mathrm{d}x)$ when $\alpha \neq 0$ and by $\int_{|x|>1}\log|x|R(\mathrm{d}x)$ when $\alpha = 0$.

We now turn to Part 2. First we show that M is, necessarily, the Lévy measure of a p-tempered α-stable distribution. From R define Q by (3.10) and note that $Q(\{0\}) = 0$. By a straightforward extension of Lemma 2.1 in [6], Q has a polar decomposition, i.e. there exists a finite Borel measure σ on $\mathbb{S}^{d-1}$ and a measurable family of Borel measures $\{Q_u\}_{u\in\mathbb{S}^{d-1}}$ on $(0,\infty)$ such that $Q(A) = \int_{\mathbb{S}^{d-1}}\int_{(0,\infty)} 1_A(ru)Q_u(\mathrm{d}r)\sigma(\mathrm{d}u)$ for $A\in\mathfrak{B}(\mathbb{R}^d)$. Define $q(s,u) := \int_{(0,\infty)} e^{-sr}Q_u(\mathrm{d}r)$ and note that (3.14) implies that for every $\alpha<2$

$$\infty > \int_{\mathbb{R}^d}\left(|x|^2\wedge|x|^\alpha\right)R(\mathrm{d}x) = \int_{\mathbb{R}^d}\left(|x|^{-(2-\alpha)/p}\wedge 1\right)Q(\mathrm{d}x)$$
$$= \int_{\mathbb{S}^{d-1}}\int_{(0,\infty)}\left(r^{-(2-\alpha)/p}\wedge 1\right)Q_u(\mathrm{d}r)\sigma(\mathrm{d}u),$$

which means that for σ a.e. u the function $q(s,u)$ is finite for every $s>0$. For $A\in\mathfrak{B}(\mathbb{R}^d)$ we have

$$\begin{aligned} M(A) &= \int_{\mathbb{R}^d}\int_0^\infty 1_A(xt)t^{-1-\alpha}e^{-t^p}\mathrm{d}tR(\mathrm{d}x)\\ &= \int_{\mathbb{R}^d}\int_0^\infty 1_A(tx|x|^{-1-1/p})t^{-1-\alpha}e^{-t^p}\mathrm{d}t|x|^{\alpha/p}Q(\mathrm{d}x)\\ &= \int_{\mathbb{S}^{d-1}}\int_{(0,\infty)}\int_0^\infty 1_A(tur^{-1/p})t^{-1-\alpha}e^{-t^p}\mathrm{d}tr^{\alpha/p}Q_u(\mathrm{d}r)\sigma(\mathrm{d}u)\\ &= \int_{\mathbb{S}^{d-1}}\int_{(0,\infty)}\int_0^\infty 1_A(us)s^{-1-\alpha}e^{-s^pr}\mathrm{d}sQ_u(\mathrm{d}r)\sigma(\mathrm{d}u)\\ &= \int_{\mathbb{S}^{d-1}}\int_0^\infty 1_A(us)q(s^p,u)s^{-1-\alpha}\mathrm{d}s\sigma(\mathrm{d}u), \end{aligned} \tag{3.16}$$

which means that this is the Lévy measure of a p-tempered α-stable distribution.

Now to show the uniqueness of R. Assume that two measures R^1 and R^2 satisfy (3.12), (3.13), and (3.14). For each $i=1,2$ define Q^i by (3.10), let $\{Q^i_u\}_{u\in\mathbb{S}^{d-1}}$ and σ^i be a polar decomposition of Q^i, and define $q^i(s,u) := \int_{(0,\infty)} e^{-sr}Q^i_u(\mathrm{d}r)$. From (3.16) it follows that we can decompose M into polar coordinates in two ways. First as $\{q^1(s^p,u)s^{-1-\alpha}\mathrm{d}s\}_{u\in\mathbb{S}^{d-1}}$ and σ^1 and second as $\{q^2(s^p,u)s^{-1-\alpha}\mathrm{d}s\}_{u\in\mathbb{S}^{d-1}}$ and σ^2. By the uniqueness of polar decompositions (see Lemma 2.1 in [6]) there exists a Borel function $c(u)$ such that $0<c(u)<\infty$,

$$\sigma^1(\mathrm{d}u) = c(u)\sigma^2(\mathrm{d}u),$$

and

$$c(u)q^1(s^p,u)s^{-1-\alpha}\mathrm{d}s = q^2(s^p,u)s^{-1-\alpha}\mathrm{d}s \text{ for } \sigma^1 \text{ a.e. } u.$$

By Theorem 16.10 in [10] and the continuity in s of $q^i(s,u)$ for $i=1,2$ this implies that for σ^1 a.e. u

$$c(u)q^1(s^p,u) = q^2(s^p,u),\quad s>0$$

which can be rewritten as

$$\int_0^\infty e^{-s^p t} c(u) Q_u^1(\mathrm{d}t) = \int_0^\infty e^{-s^p t} Q_u^2(\mathrm{d}t), \quad s > 0 \text{ for } \sigma^1 \text{ a.e. } u.$$

Since Laplace transforms uniquely determine measures we have $c(u)Q_u^1(\cdot) = Q_u^2(\cdot)$ for σ^1 a.e. u. Thus for any $A \in \mathfrak{B}(\mathbb{R}^d)$

$$\begin{aligned} Q^1(A) &= \int_{\mathbb{S}^{d-1}} \int_{(0,\infty)} 1_A(ru) Q_u^1(\mathrm{d}r) \sigma^1(\mathrm{d}u) \\ &= \int_{\mathbb{S}^{d-1}} \int_{(0,\infty)} 1_A(ru) c(u) Q_u^1(\mathrm{d}r) \frac{1}{c(u)} \sigma^1(\mathrm{d}u) \\ &= \int_{\mathbb{S}^{d-1}} \int_{(0,\infty)} 1_A(ru) Q_u^2(\mathrm{d}r) \sigma^2(\mathrm{d}u) = Q^2(A). \end{aligned}$$

By (3.9) this implies that $R^1(A) = R^2(A)$ as well.

We now consider the case of proper p-tempered α-stable distributions. Let Q be given by (3.10). From Remark 3.4 it follows that Q corresponds to a proper p-tempered α-stable distribution if and only if there is a polar decomposition of Q into $\{Q_u\}_{u \in \mathbb{S}^{d-1}}$ and σ such that Q_u is a probability measure for each $u \in \mathbb{S}^{d-1}$ and σ is a finite Borel measure on $\mathbb{S}^{d-1}$. Such a polar decomposition of Q exists if and only if Q is finite. From here the result follows by (3.11). □

Definition 3.4. Fix $\alpha < 2, p > 0$, and let $\mu \in TS_\alpha^p$. Then $\mu = ID(0, M, b)$ for some $b \in \mathbb{R}^d$ and some Lévy measure M, which can be written in the form (3.12) for a unique measure R. We call R the **Rosiński measure** of μ and we write $TS_\alpha^p(R, b)$ to denote this distribution.

An important property of p-tempered α-stable distributions is that they are closed under shifting, scaling, and convolution. Specifically, from (3.12) and (2.1) we get the following.

Proposition 3.5. *Fix $\alpha < 2$ and $p > 0$. 1. If $X \sim TS_\alpha^p(R, b)$ and $a \in \mathbb{R}$, then $aX \sim TS_\alpha^p(R_a, b_a)$, where*

$$R_a(A) = \int_{\mathbb{R}^d} 1_{A \setminus \{0\}}(ax) R(\mathrm{d}x), \qquad A \in (\mathbb{R}^d)$$

and

$$\begin{aligned} b_a &= ab + \int_{\mathbb{R}^d} \int_0^\infty \left(\frac{ax}{1 + a^2 t^2 |x|^2} - \frac{ax}{1 + t^2 |x|^2} \right) t^{-\alpha} e^{-t^p} \mathrm{d}t R(\mathrm{d}x) \\ &= ab + a(1 - a^2) \int_{\mathbb{R}^d} \int_0^\infty \frac{x|x|^2}{(1 + a^2 t^2 |x|^2)(1 + t^2 |x|^2}) t^{2-\alpha} e^{-t^p} \mathrm{d}t R(\mathrm{d}x). \end{aligned}$$

2. *If* $X_1 \sim TS^p_\alpha(R_1, b_1)$ *and* $X_2 \sim TS^p_\alpha(R_2, b_2)$ *are independent and* $b \in \mathbb{R}^d$, *then*

$$X_1 + X_2 + b \sim TS^p_\alpha(R_1 + R_2, b_1 + b_2 + b),$$

where $R_1 + R_2$ *is the Borel measure defined by* $(R_1 + R_2)(B) = R_1(B) + R_2(B)$ *for any* $B \in \mathfrak{B}(\mathbb{R}^d)$.

For proper p-tempered α-stable distributions we can recover the representation of the Lévy measure given by (3.2) as follows.

Proposition 3.6. *Fix* $\alpha < 2$, $p > 0$, *and let* M *be the Lévy measure of a proper* p*-tempered* α*-stable distribution with Rosiński measure* R. M *can be represented by* (3.2) *with* $q(r, u)$ *satisfying* (3.7) *and*

$$\sigma(B) = \int_{\mathbb{R}^d} 1_B\left(\frac{x}{|x|}\right)|x|^\alpha R(\mathrm{d}x), \qquad B \in \mathfrak{B}(\mathbb{S}^{d-1}). \tag{3.17}$$

If, in addition, $\alpha \in (0, 2)$, *then the Lévy measure of an* α*-stable distribution with spectral measure* σ *is given by*

$$L(B) = \int_{\mathbb{R}^d}\int_0^\infty 1_B(tx)\, t^{-\alpha-1}\mathrm{d}t R(\mathrm{d}x), \qquad B \in \mathfrak{B}(\mathbb{R}^d).$$

Proof. Let Q be derived from R by (3.10). Remark 3.4 implies that there is a finite Borel measure σ on $\mathbb{S}^{d-1}$ and a measurable family of probability measures $\{Q_u\}_{u\in\mathbb{S}^{d-1}}$ such that Q can be represented in terms of σ and $\{Q_u\}_{u\in\mathbb{S}^{d-1}}$ as in (3.8) and that M can be represented by (3.2) where $q(r, u) = \int_{(0,\infty)} e^{-sr} Q_u(\mathrm{d}r)$. From here it follows that $q(r, u)$ satisfies (3.7) by the Monotone Convergence Theorem and the fact that Q_u is a probability measure for each $u \in \mathbb{S}^{d-1}$. Further, for any $A \in \mathfrak{B}(\mathbb{S}^{d-1})$

$$\begin{aligned}\int_{\mathbb{R}^d} 1_A\left(\frac{x}{|x|}\right)|x|^\alpha R(\mathrm{d}x) &= \int_{\mathbb{R}^d} 1_A\left(\frac{x}{|x|}\right) Q(\mathrm{d}x)\\ &= \int_A\int_{(0,\infty)} Q_u(\mathrm{d}s)\sigma(\mathrm{d}u) = \sigma(B).\end{aligned}$$

The second part follows from the first and the fact that for any $A \in \mathfrak{B}(\mathbb{R}^d)$

$$\begin{aligned}L(A) &= \int_{\mathbb{S}^{d-1}}\int_0^\infty 1_B(su)s^{-\alpha-1}\mathrm{d}s\sigma(\mathrm{d}u)\\ &= \int_{\mathbb{R}^d}\int_0^\infty 1_B(sx/|x|)s^{-1-\alpha}\mathrm{d}s|x|^\alpha R(\mathrm{d}x)\\ &= \int_{\mathbb{R}^d}\int_0^\infty 1_B(tx)t^{-\alpha-1}\mathrm{d}tR(\mathrm{d}x),\end{aligned}$$

where the third equality follows by the substitution $t = s/|x|$. □

3.2 Identifiability and Subclasses

In Theorem 3.3 we saw that for fixed $p > 0$ and $\alpha < 2$ there is a one-to-one relationship between the Rosiński measure R and the Lévy measure M. We may further ask whether all of the parameters are jointly identifiable. Unfortunately, the answer is negative. In fact, even for fixed $p > 0$, the parameters α and R are not jointly identifiable. However, for fixed $p > 0$, in the subclass of proper tempered stable distribution, they are jointly identifiable. On the other hand, for fixed $\alpha < 2$, even in the subclass of proper tempered stable distributions, the parameters p and R are not jointly identifiable. These facts will be verified in this section. We begin with a lemma.

Lemma 3.7. *Fix $\alpha < 2$, $p > 0$, and let M be the Lévy measure of a p-tempered α-stable distribution with Rosiński measure $R \neq 0$.*

1. *The map $s \mapsto s^\alpha M(|x| > s)$ is decreasing and $\lim_{s\to\infty} s^\alpha M(|x| > s) = 0$.*
2. *If $\alpha \in (0, 2)$, then*

$$\lim_{s\downarrow 0} s^\alpha M(|x| > s) = \frac{1}{\alpha}\int_{\mathbb{R}^d} |x|^\alpha R(\mathrm{d}x)$$

and if $\alpha \leq 0$, then

$$\lim_{s\downarrow 0} s^\alpha M(|x| > s) = \infty.$$

3. *If $\alpha < 0$, then*

$$\lim_{s\downarrow 0} s^\alpha M(|x| < s) = \frac{1}{|\alpha|}\int_{\mathbb{R}^d} |x|^\alpha R(\mathrm{d}x)$$

and if $\alpha \in [0, 2)$, then for all $s > 0$

$$M(|x| < s) = \infty.$$

Proof. We begin with the first part. Since

$$\begin{aligned} s^\alpha M(|x| > s) &= s^\alpha \int_{\mathbb{R}^d}\int_{s|x|^{-1}}^\infty t^{-1-\alpha} e^{-t^p} \mathrm{d}t R(\mathrm{d}x) \\ &= \int_{\mathbb{R}^d}\int_{|x|^{-1}}^\infty t^{-1-\alpha} e^{-(st)^p} \mathrm{d}t R(\mathrm{d}x), \end{aligned} \tag{3.18}$$

the map $s \mapsto s^\alpha M(|x| > s)$ is decreasing. For large enough s, the integrand in (3.18) is bounded by $1_{t>1/|x|} t^{-1-\alpha} e^{-t^p}$, which is integrable. Thus by dominated convergence $\lim_{s\to\infty} s^\alpha M(|x| > s) = 0$.

For the second part, by (3.18) and the Monotone Convergence Theorem

$$\lim_{s\downarrow 0} s^{\alpha} M(|x| > s) = \int_{\mathbb{R}^d} \int_{|x|^{-1}}^{\infty} t^{-1-\alpha} \mathrm{d}t R(\mathrm{d}x).$$

Thus if $\alpha \in (0, 2)$, then

$$\lim_{s\downarrow 0} s^{\alpha} M(|x| > s) = \frac{1}{\alpha} \int_{\mathbb{R}^d} |x|^{\alpha} R(\mathrm{d}x),$$

and if $\alpha \le 0$, then

$$\lim_{s\downarrow 0} s^{\alpha} M(|x| > s) = \infty.$$

We now show the third part. If $\alpha \in [0, 2)$, then for all $s > 0$

$$\begin{aligned} M(|x| < s) &= \int_{\mathbb{R}^d} \int_0^{s|x|^{-1}} t^{-1-\alpha} e^{-t^p} \mathrm{d}t R(\mathrm{d}x) \\ &\ge \int_{\mathbb{R}^d} e^{-(s/|x|)^p} \int_0^{s|x|^{-1}} t^{-1-\alpha} \mathrm{d}t R(\mathrm{d}x) = \infty, \end{aligned}$$

and if $\alpha < 0$, then

$$\begin{aligned} \lim_{s\downarrow 0} s^{\alpha} M(|x| < s) &= \lim_{s\downarrow 0} s^{\alpha} \int_{\mathbb{R}^d} \int_0^{s|x|^{-1}} t^{-1-\alpha} e^{-t^p} \mathrm{d}t R(\mathrm{d}x) \\ &= \lim_{s\downarrow 0} \int_{\mathbb{R}^d} \int_0^{|x|^{-1}} t^{-1-\alpha} e^{-(st)^p} \mathrm{d}t R(\mathrm{d}x) \\ &= \int_{\mathbb{R}^d} \int_0^{|x|^{-1}} t^{-1-\alpha} \mathrm{d}t R(\mathrm{d}x) = \frac{1}{|\alpha|} \int_{\mathbb{R}^d} |x|^{\alpha} R(\mathrm{d}x), \end{aligned}$$

where the third line follows by the Monotone Convergence Theorem. □

Combining Lemma 3.7 with (3.15) gives the following.

Proposition 3.8. *In the subclass of proper tempered stable distributions with parameter $p > 0$ fixed, the parameters R and α are jointly identifiable.*

However, in general, the parameters α and p are not identifiable. This will become apparent from the following results.

Proposition 3.9. *Fix $\alpha < 2$, $\beta \in (\alpha, 2)$, and let $K = \int_0^{\infty} s^{\beta-\alpha-1} e^{-s^p} \mathrm{d}s$. If $\mu = TS_{\beta}^{p}(R, b)$ and*

$$R'(A) = K^{-1} \int_{\mathbb{R}^d} \int_0^1 1_A(ux) u^{-\beta-1} (1-u^p)^{(\beta-\alpha)/p-1} \mathrm{d}u R(\mathrm{d}x),$$

then R' is the Rosiński measure of a p-tempered α-stable distribution and $\mu = TS_\alpha^p(R', b)$.

Proof. We begin by verifying that R' is the Rosiński measure of some p-tempered α-stable distribution. Let $C = \max_{u\in[0,.5]} (1-u^p)^{(\beta-\alpha)/p-1}$. We have

$$\begin{aligned}
K\int_{|x|\le 1} |x|^2 R'(\mathrm{d}x) &= \int_{\mathbb{R}^d} |x|^2 \int_0^{1\wedge|x|^{-1}} u^{1-\beta}(1-u^p)^{(\beta-\alpha)/p-1}\mathrm{d}u R(\mathrm{d}x)\\
&\le \int_{|x|\le 2} |x|^2 R(\mathrm{d}x) \int_0^1 u^{1-\beta}(1-u^p)^{(\beta-\alpha)/p-1}\mathrm{d}u\\
&\quad + C\int_{|x|>2} |x|^2 \int_0^{|x|^{-1}} u^{1-\beta}\mathrm{d}u R(\mathrm{d}x)\\
&= \int_{|x|\le 2} |x|^2 R(\mathrm{d}x) \int_0^1 u^{1-\beta}(1-u^p)^{(\beta-\alpha)/p-1}\mathrm{d}u\\
&\quad + \frac{C}{2-\beta}\int_{|x|\ge 2} |x|^\beta R(\mathrm{d}x) < \infty.
\end{aligned}$$

If $\alpha \in (0,2)$, then

$$\begin{aligned}
K\int_{|x|>1} |x|^\alpha R'(\mathrm{d}x) &= \int_{|x|\ge 1} |x|^\alpha \int_{|x|^{-1}}^1 u^{\alpha-\beta-1}(1-u^p)^{(\beta-\alpha)/p-1}\mathrm{d}u R(\mathrm{d}x)\\
&\le \int_{|x|\ge 2} |x|^\alpha \int_{|x|^{-1}}^{1/2} u^{\alpha-\beta-1}(1-u^p)^{(\beta-\alpha)/p-1}\mathrm{d}u R(\mathrm{d}x)\\
&\quad + \int_{|x|\ge 1} |x|^\alpha \int_{1/2}^1 u^{\alpha-\beta-1}(1-u^p)^{(\beta-\alpha)/p-1}\mathrm{d}u R(\mathrm{d}x)\\
&\le C\int_{|x|\ge 2} |x|^\alpha \int_{|x|^{-1}}^\infty u^{\alpha-\beta-1}\mathrm{d}u R(\mathrm{d}x)\\
&\quad + \int_{|x|\ge 1} |x|^\beta R(\mathrm{d}x) \int_{1/2}^1 u^{\alpha-\beta-1}(1-u^p)^{(\beta-\alpha)/p-1}\mathrm{d}u,
\end{aligned}$$

which is finite since the first integral equals $\frac{C}{\beta-\alpha}\int_{|x|\ge 2} |x|^\beta R(\mathrm{d}x) < \infty$. Now assume $\alpha = 0$ and fix $\epsilon \in (0,\beta)$. By 4.1.37 in [2] there exists a $C_\epsilon > 0$ such that for all $u > 0$, $\log u \le C_\epsilon u^\epsilon$. Thus

$$K\int_{|x|>1} \log|x| R'(\mathrm{d}x) \le KC_\epsilon \int_{|x|>1} |x|^\epsilon R'(\mathrm{d}x),$$

which is finite by arguments similar to the previous case. When $\alpha < 0$

$$K\int_{|x|>1} R'(\mathrm{d}x) = \int_{|x|\geq 1}\int_{|x|^{-1}}^{1} u^{-\beta-1}(1-u^p)^{(\beta-\alpha)/p-1}\mathrm{d}uR(\mathrm{d}x)$$
$$\leq C\int_{|x|\geq 2}\int_{|x|^{-1}}^{1} u^{-\beta-1}\mathrm{d}uR(\mathrm{d}x)$$
$$+\int_{|x|\geq 1} R(\mathrm{d}x)\int_{1/2}^{1} u^{-\beta-1}(1-u^p)^{(\beta-\alpha)/p-1}\mathrm{d}u,$$

which is finite since for $\beta \neq 0$ the first integral is $\frac{C}{\beta}\int_{|x|>2}\left(|x|^{\beta}-1\right)R(\mathrm{d}x) < \infty$ and for $\beta = 0$ it is $\int_{|x|>2}\log|x|R(\mathrm{d}x) < \infty$. Now, let M' be the Lévy measure of $TS^p_\alpha(R', b)$. By (3.12) for $A \in \mathfrak{B}(\mathbb{R}^d)$ we have

$$M'(A) = K^{-1}\int_{\mathbb{R}^d}\int_0^\infty\int_0^1 1_A(utx)t^{-1-\alpha}e^{-t^p}u^{-\beta-1}\left(1-u^p\right)^{\frac{\beta-\alpha}{p}-1}\mathrm{d}u\mathrm{d}tR(\mathrm{d}x)$$
$$= K^{-1}\int_{\mathbb{R}^d}\int_0^\infty\int_0^t 1_A(vx)t^{\beta-\alpha-1}e^{-t^p}v^{-\beta-1}\left(1-v^p/t^p\right)^{\frac{\beta-\alpha}{p}-1}\mathrm{d}v\mathrm{d}tR(\mathrm{d}x)$$
$$= K^{-1}\int_{\mathbb{R}^d}\int_0^\infty\int_v^\infty 1_A(vx)t^{p-1}e^{-t^p}v^{-\beta-1}\left(t^p-v^p\right)^{\frac{\beta-\alpha}{p}-1}\mathrm{d}t\mathrm{d}vR(\mathrm{d}x)$$
$$= K^{-1}\int_{\mathbb{R}^d}\int_0^\infty 1_A(vx)e^{-v^p}v^{-\beta-1}\mathrm{d}vR(\mathrm{d}x)\int_0^\infty e^{-s^p}s^{\beta-\alpha-1}\mathrm{d}s$$
$$= \int_{\mathbb{R}^d}\int_0^\infty 1_A(vx)e^{-v^p}v^{-\beta-1}\mathrm{d}vR(\mathrm{d}x),$$

where the second line follows by the substitution $v = ut$ and the fourth by the substitution $s^p = t^p - v^p$. □

To show a similar result for the parameter p we need some additional notation. For $r \in (0, 1)$, let f_r be a probability density with $f_r(x) = 0$ for $x < 0$ and

$$\int_0^\infty e^{-tx}f_r(x)\mathrm{d}x = e^{-t^r}.$$

Such a density exists, and is, in fact, the density of a certain type of r-stable distribution, see Proposition 1.2.12 in [68]. The only case where an explicit formula is known is

$$f_{.5}(s) = \frac{1}{2\sqrt{\pi}}e^{-1/(4s)}s^{-3/2}1_{[s>0]}$$

(see Examples 2.13 and 8.11 in [69]). From Theorem 5.4.1 in [78] it follows that if $\beta \geq 0$, then

$$\int_0^\infty s^{-\beta}f_r(s)\mathrm{d}s < \infty.$$

Proposition 3.10. *Fix $\alpha < 2$ and $0 < p < q$. If $\mu = TS^p_\alpha(R, b)$ and*

$$R'(A) = \int_{\mathbb{R}^d} \int_0^\infty 1_A(s^{-1/q}x) s^{\alpha/q} f_{p/q}(s) \mathrm{d}s R(\mathrm{d}x),$$

then R' is the Rosiński measure of a q-tempered α-stable distribution and $\mu = TS^q_\alpha(R', b)$. Moreover, μ is a proper p-tempered α-stable distribution if and only if it is a proper q-tempered α-stable distribution.

This implies that, for fixed α, the parameters p and R are not jointly identifiable even within the subclass of proper tempered stable distributions.

Proof. We begin by verifying that R' is, in fact, the Rosiński measure of a q-tempered α-stable distribution. We have

$$\begin{aligned}\int_{|x|\le 1} |x|^2 R'(\mathrm{d}x) &= \int_{\mathbb{R}^d} |x|^2 \int_{|x|^q}^\infty s^{-(2-\alpha)/q} f_{p/q}(s) \mathrm{d}s R(\mathrm{d}x) \\ &\le \int_{|x|\le 1} |x|^2 \int_0^\infty s^{-(2-\alpha)/q} f_{p/q}(s) \mathrm{d}s R(\mathrm{d}x) \\ &\quad + \int_{|x|>1} |x|^\alpha R(\mathrm{d}x) \int_0^\infty f_{p/q}(s) \mathrm{d}s < \infty.\end{aligned}$$

If $\alpha \neq 0$ and $\beta = \alpha \vee 0$, then

$$\begin{aligned}\int_{|x|>1} |x|^\beta R'(\mathrm{d}x) &= \int_{\mathbb{R}^d} |x|^\beta \int_0^{|x|^q} s^{-(\beta-\alpha)/q} f_{p/q}(s) \mathrm{d}s R(\mathrm{d}x) \\ &\le \int_{|x|\le 1} |x|^2 \int_0^\infty s^{-(2-\alpha)/q} f_{p/q}(s) \mathrm{d}s R(\mathrm{d}x) \\ &\quad + \int_{|x|>1} |x|^\beta \int_0^\infty s^{-(\beta-\alpha)/q} f_{p/q}(s) \mathrm{d}s R(\mathrm{d}x) < \infty.\end{aligned}$$

If $\alpha = 0$, then

$$\begin{aligned}\int_{|x|>1} \log|x| R'(\mathrm{d}x) &= \int_{\mathbb{R}^d} \int_0^{|x|^q} \log|xs^{-1/q}| f_{p/q}(s) \mathrm{d}s R(\mathrm{d}x) \\ &\le .5 \int_{|x|\le 1} |x|^2 R(\mathrm{d}x) \int_0^\infty s^{-2/q} f_{p/q}(s) \mathrm{d}s \\ &\quad + \int_{|x|>1} \log|x| R(\mathrm{d}x) \int_0^\infty f_{p/q}(s) \mathrm{d}s \\ &\quad + \int_{|x|>1} R(\mathrm{d}x) \int_0^\infty s^{-1/q} f_{p/q}(s) \mathrm{d}s < \infty,\end{aligned}$$

where the inequality uses the fact that $\log|x| \le |x|$ (see 4.1.36 in [2]).

If M' is the Lévy measure of $TS_\alpha^q(R', b)$, then by (3.12) for any $A \in \mathfrak{B}(\mathbb{R}^d)$

$$\begin{aligned} M'(A) &= \int_{\mathbb{R}^d}\int_0^\infty\int_0^\infty 1_A(s^{-1/q}tx)t^{-1-\alpha}e^{-t^q}\mathrm{d}ts^{\alpha/q}f_{p/q}(s)\mathrm{d}sR(\mathrm{d}x)\\ &= \int_{\mathbb{R}^d}\int_0^\infty 1_A(vx)v^{-1-\alpha}\int_0^\infty e^{-v^q s}f_{p/q}(s)\mathrm{d}s\mathrm{d}vR(\mathrm{d}x)\\ &= \int_{\mathbb{R}^d}\int_0^\infty 1_A(vx)v^{-1-\alpha}e^{-v^p}\mathrm{d}vR(\mathrm{d}x), \end{aligned}$$

where $v = s^{-1/q}t$. The last part follows from (3.15) and the fact that

$$\int_{\mathbb{R}^d}|x|^\alpha R'(\mathrm{d}x) = \int_{\mathbb{R}^d}|x|^\alpha R(\mathrm{d}x)\int_0^\infty s^{-\alpha/q}s^{\alpha/q}f_{p/q}(s)\mathrm{d}s = \int_{\mathbb{R}^d}|x|^\alpha R(\mathrm{d}x).$$

This concludes the proof. □

Propositions 3.9 and 3.10 give a constructive proof of the following.

Proposition 3.11. *Fix $\alpha < 2$, $p > 0$, and let $\mu \in TS_\alpha^p$.*

1. *For any $q \geq p$, $\mu \in TS_\alpha^q$.*
2. *For any $\beta \leq \alpha$, $\mu \in TS_\beta^p$.*

We now characterize when a p-tempered α-stable distribution is β-stable for some $\beta \in (0, 2)$.

Proposition 3.12. *Fix $\alpha < 2$, $p > 0$, and $\beta \in (0, 2)$. Let $\mu = S_\beta(\sigma, b)$, where $\sigma \neq 0$. If $\beta \leq \alpha$, then $\mu \notin TS_\alpha^p$. If $\beta \in (0 \vee \alpha, 2)$, then $\mu = TS_\alpha^p(R_\sigma^\beta, b)$ and*

$$R_\sigma^\beta(A) = K^{-1}\int_{\mathbb{S}^{d-1}}\int_0^\infty 1_A(ru)r^{-1-\beta}\mathrm{d}r\sigma(\mathrm{d}u), \quad A \in \mathfrak{B}(\mathbb{R}^d), \tag{3.19}$$

where $K = \int_0^\infty t^{\beta-\alpha-1}e^{-t^p}\mathrm{d}t$.

Combining (3.15) with the fact that

$$\int_{\mathbb{R}^d}|x|^\alpha R_\sigma^\beta(\mathrm{d}x) = K^{-1}\sigma(\mathbb{S}^{d-1})\int_0^\infty r^{-(\beta-\alpha)-1}\mathrm{d}r = \infty,$$

shows that no stable distributions belong to the subclass of proper p-tempered α-stable distributions.

Proof. If $\mu \in TS_\alpha^p$, then its Lévy measure can be written as (3.2). By uniqueness of the polar decomposition of Lévy measures (see Lemma 2.1 in [6]) there exists a nonnegative Borel function $c(u)$ with $\sigma(\{u : c(u) > 0\}) > 0$ such that $q(r, u) = c(u)r^{(\alpha-\beta)/p}$. This does not satisfy (3.4) when $\beta \leq \alpha$.

Now assume that $\beta > \alpha$. In this case $R_\sigma^\beta(\{0\}) = 0$ and for any $\gamma \in [0, \beta)$

$$\int_{\mathbb{R}^d} \left(|x|^2 \wedge |x|^\gamma\right) R_\sigma^\beta(\mathrm{d}x) = K^{-1}\sigma(\mathbb{S}^{d-1}) \int_0^\infty (r^{1-\beta} \wedge r^{\gamma-\beta-1})\mathrm{d}r < \infty.$$

Thus, by Theorem 3.3, R_σ^β is the Rosiński measure of a p-tempered α-stable distribution. If M is the Lévy measure of $TS_\alpha^p(R_\sigma^\beta, b)$, then for any $A \in \mathfrak{B}(\mathbb{R}^d)$

$$\begin{aligned} M(A) &= K^{-1} \int_{\mathbb{S}^{d-1}} \int_0^\infty \int_0^\infty 1_A(rtu) t^{-1-\alpha} e^{-t^p} \mathrm{d}t r^{-1-\beta} \mathrm{d}r \sigma(du) \\ &= K^{-1} \int_0^\infty t^{\beta-\alpha-1} e^{-t^p} \mathrm{d}t \int_{\mathbb{S}^{d-1}} \int_0^\infty 1_A(ru) r^{-1-\beta} \mathrm{d}r \sigma(\mathrm{d}u) \\ &= \int_{\mathbb{S}^{d-1}} \int_0^\infty 1_A(ru) r^{-1-\beta} \mathrm{d}r \sigma(\mathrm{d}u), \end{aligned}$$

which is the Lévy measure of μ. □

Recall that a probability measure μ is called compound Poisson if its characteristic function can be written as

$$\hat{\mu}(z) = \exp\left\{ \int_{\mathbb{R}^d} \left(e^{i\langle z,x\rangle} - 1\right) M(\mathrm{d}x) \right\}, \qquad z \in \mathbb{R}^d,$$

where M is a finite Lévy measure. To classify when tempered stable distributions are compound Poisson we begin with a lemma.

Lemma 3.13. *Let M be given by* (3.12). *M is finite if and only if either $R = 0$ or $\alpha < 0$ and R is a finite measure.*

Proof. Observing that

$$\begin{aligned} R(\mathbb{R}^d)e^{-1} \int_0^1 t^{-1-\alpha} \mathrm{d}t &\le \int_{\mathbb{R}^d} \int_0^\infty e^{-t^p} t^{-1-\alpha} \mathrm{d}t R(\mathrm{d}x) \\ &\le R(\mathbb{R}^d) \left(\int_0^1 t^{-1-\alpha} \mathrm{d}t + \int_1^\infty e^{-t^p} t^{-1-\alpha} \mathrm{d}t \right) \end{aligned}$$

gives the result. □

This immediately gives the following.

Proposition 3.14. *If $\mu = TS_\alpha^p(R, b)$, then μ is compound Poisson if and only if either $R = 0$ or $\alpha < 0$, R is a finite measure, and $b = \int_{\mathbb{R}^d} \int_0^\infty \frac{x}{1+t^2|x|^2} t^{-\alpha} e^{-t^p} \mathrm{d}t R(\mathrm{d}x)$.*

3.3 Tails of Tempered Stable Distributions

Since the motivation for introducing p-tempered α-stable distributions is to get models with tails lighter than those of α-stable distributions, it is important to understand how the tails behave. One of the easiest ways to describe the tails of a distribution is to characterize which moments are finite. Toward this end we present several results that were proved in [27]. Throughout this section we adopt the convention that $0^0 = 1$.

Theorem 3.15. *Fix $\alpha < 2$, $p > 0$, and let $\mu = TS^p_\alpha(R, b)$.*

1. *If $\alpha \in (0, 2)$ and $q_1, \ldots, q_d \geq 0$ with $q := \sum_{j=1}^d q_j < \alpha$, then*

$$\int_{\mathbb{R}^d} \left(\prod_{j=1}^d |x_j|^{q_j} \right) \mu(\mathrm{d}x) \leq \int_{\mathbb{R}^d} |x|^q \mu(\mathrm{d}x) < \infty.$$

2. *If $\alpha \in (0, 2)$, then*

$$\int_{\mathbb{R}^d} |x|^\alpha \mu(\mathrm{d}x) < \infty \Longleftrightarrow \int_{|x|>1} |x|^\alpha \log |x| R(\mathrm{d}x) < \infty.$$

Additionally, if $q_1, \ldots, q_d \geq 0$ with $\sum_{j=1}^d q_j = \alpha$, then

$$\int_{\mathbb{R}^d} \left(\prod_{j=1}^d |x_j|^{q_j} \right) \mu(\mathrm{d}x) < \infty$$

if and only if

$$\int_{|x|>1} \left(\prod_{j=1}^d |x_j|^{q_j} \right) \log |x| R(\mathrm{d}x) < \infty. \tag{3.20}$$

3. *If $q > (\alpha \vee 0)$, then*

$$\int_{\mathbb{R}^d} |x|^q \mu(\mathrm{d}x) < \infty \Longleftrightarrow \int_{|x|>1} |x|^q R(\mathrm{d}x) < \infty.$$

Additionally, if $q_1, \ldots, q_d \geq 0$ with $\sum_{j=1}^d q_j > (\alpha \vee 0)$, then

$$\int_{\mathbb{R}^d} \left(\prod_{j=1}^d |x_j|^{r_j} \right) \mu(\mathrm{d}x) < \infty \text{ for all } r_k \in [0, q_k],\ k = 1, \ldots, d$$

if and only if

$$\int_{|x|>1}\left(\prod_{j=1}^{d}|x_j|^{r_j}\right)R(\mathrm{d}x)<\infty \textit{ for all } r_k\in[0,q_k],\ k=1,\dots,d. \quad (3.21)$$

Further, we can find explicit formulas for the moments and the mixed moments. However, these formulas can get quite complicated. When working with infinitely divisible distribution it is often easier to find the cumulants instead. Recall that for any infinitely divisible distribution μ the function C_μ given by (2.1) is called the cumulant generating function. This name is explained by the following. Let $k=(k_1,k_2,\dots,k_d)$ be a d-dimensional vector of nonnegative integers and let

$$c_k=(-i)^{\sum k_i}\frac{\partial^{\sum k_i}}{\partial z_d^{k_d}\cdots\partial z_1^{k_1}}C_\mu(z)\bigg|_{z=0},$$

whenever the derivative exists and is continuous in a neighborhood of zero. We call this the cumulant of order k. The cumulants can be uniquely expressed in terms of the moments, see, e.g., [73]. In particular let $X\sim\mu$. When $k_i=1$ and $k_j=0$ for all $j\neq i$ then $c_k=\mathrm{E}[X_i]$, when $k_i=2$ and $k_j=0$ for all $j\neq i$ then $c_k=\mathrm{var}(X_i)$, and when for some $i\neq j$ we have $k_i=k_j=1$ and $k_\ell=0$ for all $\ell\neq i,j$ then $c_k=\mathrm{cov}(X_i,X_j)$. The following is given in [27].

Theorem 3.16. *Fix $\alpha<2$, $p>0$, and let $\mu=TS^p_\alpha(R,b)$. Let $q_1,\dots,q_d$ be nonnegative integers and let $q_+=\sum_{i=1}^d q_i$. Further, if $q_+=\alpha=1$, assume that* (3.20) *holds and if $q_+>\alpha$, that* (3.21) *holds. If $q_i=q_+=1$ for some i, then*

$$c_{(q_1,\dots,q_d)}=b_i+\int_{\mathbb{R}^d}\int_0^\infty x_i\frac{|x|^2}{1+|x|^2t^2}t^{2-\alpha}e^{-t^p}\mathrm{d}tR(\mathrm{d}x).$$

If $q_+\geq 2$, then

$$c_{(q_1,\dots,q_d)}=p^{-1}\Gamma\left(\frac{q_+-\alpha}{p}\right)\int_{\mathbb{R}^d}\left(\prod_{j=1}^d x_j^{q_j}\right)R(\mathrm{d}x).$$

We now turn to the question of exponential moments.

Theorem 3.17. *Fix $\alpha<2$, $p\in(0,1]$, and $\theta>0$. Let $\mu=TS^p_\alpha(R,b)$.*

1. If $\alpha\in(0,2)$, then $\int_{\mathbb{R}^d}e^{\theta|x|^p}\mu(\mathrm{d}x)<\infty$ if and only if

$$R(|x|>\theta^{-1/p})=0.$$

2. If $\alpha=0$, then $\int_{\mathbb{R}^d}e^{\theta|x|^p}\mu(\mathrm{d}x)<\infty$ if and only if

$$R(|x|\geq\theta^{-1/p})=0 \textit{ and } \int_{0<|x|^{-p}-\theta<1}|\log(|x|^{-p}-\theta)|\,R(\mathrm{d}x)<\infty.$$

3. *If* $\alpha < 0$, *then* $\int_{\mathbb{R}^d} e^{\theta|x|^p}\mu(\mathrm{d}x) < \infty$ *if and only if*

$$R(|x| \geq \theta^{-1/p}) = 0 \text{ and } \int_{0<|x|^{-p}-\theta<1} (|x|^{-p} - \theta)^{\alpha/p} R(\mathrm{d}x) < \infty.$$

Further, from Theorem 4 in [27] it follows that if $p > 1$ and there exists an $\epsilon > 0$ such that

$$\int_{|x|>1} e^{|x|^{\epsilon+p/(p-1)}} |x|^{-\alpha/(p-1)} R(\mathrm{d}x) < \infty, \tag{3.22}$$

then

$$\int_{\mathbb{R}^d} e^{\theta|x|}\mu(\mathrm{d}x) < \infty \text{ for all } \theta \geq 0. \tag{3.23}$$

However, the tails cannot be too light and if $R \neq 0$, then

$$\int_{\mathbb{R}^d} e^{\theta|x|\log|x|}\mu(\mathrm{d}x) = \infty \text{ for all } \theta > 0.$$

When the exponential moments exist, we can evaluate them. Specifically, if $\mu = TS^p_\alpha(R,b)$ and $z \in \mathbb{C}^d$ is such that $\int_{\mathbb{R}^d} e^{\langle x, \Re z\rangle}\mu(\mathrm{d}x) < \infty$, then Theorem 25.17 in [69] implies that $\int_{\mathbb{R}^d} |e^{\langle x,z\rangle}|\mu(\mathrm{d}x) < \infty$ and that $\int_{\mathbb{R}^d} e^{\langle x,z\rangle}\mu(\mathrm{d}x)$ is given by

$$\exp\left\{\int_{\mathbb{R}^d}\int_0^\infty \left(e^{\langle x,z\rangle t} - 1 - \frac{t\langle x,z\rangle}{1+|x|^2}\right) t^{-1-\alpha} e^{-t^p} \mathrm{d}t R(\mathrm{d}x) + \langle z,b\rangle\right\}. \tag{3.24}$$

For the case $p = 1$ more explicit formulas will be given in Section 3.5.

Another way to analyze the tails of a probability measure is to ask when they are regularly varying. First consider the case where $\mu = TS^p_\alpha(R,b)$ with $\alpha \in (0,2)$. Theorem 3.15 implies that $\int_{\mathbb{R}^d} |x|^\eta\mu(\mathrm{d}x) < \infty$ for all $\eta \in [0,\alpha)$, and hence, by Proposition 2.12, μ cannot have regularly varying tails with tail index $|\gamma| < \alpha$. However, other tail indices are possible. The following result from [27] characterizes this.

Theorem 3.18. *Fix* $\alpha < 2$ *and* $p > 0$. *Let* $\mu = TS^p_\alpha(R,b)$ *and let* M *be the Lévy measure of* μ. *If* $\gamma < (-\alpha) \wedge 0$, *then*

$$\mu \in RV^\infty_\gamma(\sigma) \iff M \in RV^\infty_\gamma(\sigma) \iff R \in RV^\infty_\gamma(\sigma).$$

Moreover, if $M \in RV^\infty_\gamma(\sigma)$, *then for all* $D \in \mathfrak{B}(\mathbb{S}^{d-1})$ *with* $\sigma(\partial D) = 0$ *and* $\sigma(D) > 0$

$$\lim_{r\to\infty} \frac{R(|x| > r, x/|x| \in D)}{M(|x| > r, x/|x| \in D)} = \frac{p}{\Gamma\left(\frac{|\gamma|-\alpha}{p}\right)}.$$

Now recall that for $\beta \in (0,2)$ a probability measure μ belongs to the domain of attraction of a β-stable distribution with spectral measure $\sigma \neq 0$ if and only if $\mu \in RV^{\infty}_{-\beta}(\sigma)$. See, e.g., [67] or [54] although they make the additional assumption that the limiting stable distribution is full. This leads to the following.

Corollary 3.19. *Fix $\alpha < 2$, $p > 0$, let $\mu = TS^p_\alpha(R,b)$, and let $\sigma \neq 0$ be a finite Borel measure on $\mathbb{S}^{d-1}$. If $\beta \in (0 \vee \alpha, 2)$, then μ belongs to the domain of attraction of a β-stable distribution with spectral measure σ if and only if $R \in RV^{\infty}_{-\beta}(\sigma)$.*

3.4 Tempered Stable Lévy Processes

Fix $\alpha < 2$ and $p > 0$. A Lévy process $\{X_t : t \geq 0\}$ is called a **p-tempered α-stable Lévy process** if $X_1 \sim TS^p_\alpha(R,b)$. In this section we discuss properties of such processes.

Proposition 3.20. *Let $\{X_t : t \geq 0\}$ be a Lévy process with $X_1 \sim TS^p_\alpha(R,b)$, and assume that $R \neq 0$.*

1. *The paths of $\{X_t : t \geq 0\}$ are discontinuous a.s.*
2. *The paths of $\{X_t : t \geq 0\}$ are piecewise constant a.s. if and only if $\alpha < 0$, R is a finite measure, and $b = \int_{\mathbb{R}^d}\int_0^\infty \frac{x}{1+t^2|x|^2} t^{-\alpha} e^{-t^p} \mathrm{d}t R(\mathrm{d}x)$.*
3. *If $\alpha < 0$ and R is a finite measure, then, almost surely, jumping times are infinitely many and countable in increasing order. The first jumping time has an exponential distribution with mean $1/a$, where $a = R(\mathbb{R}^d)p^{-1}\Gamma(|\alpha|/p)$.*
4. *If $\alpha \geq 0$ or R is an infinite measure, then, almost surely, jumping times are countable and dense in $[0,\infty)$.*

Proof. Part 1 follows by Theorem 21.1 in [69]. Part 2 follows by Theorem 21.2 in [69] and Proposition 3.14. Parts 3 and 4 follow by Theorem 21.3 in [69] and Lemma 3.13. □

A useful index that determines many properties of Lévy processes was introduced by Blumenthal and Getoor [12]. It is defined as follows.

Definition 3.21. Let $\{X_t : t \geq 0\}$ be a Lévy process with $X_1 \sim ID(0,M,b)$. The number

$$\beta = \inf\left\{\gamma > 0 : \int_{|x|\leq 1} |x|^\gamma M(\mathrm{d}x) < \infty\right\}$$

is called the **Blumenthal-Getoor index**.

From the definition of a Lévy measure, it is clear that the Blumenthal-Getoor index is a number in $[0,2]$.

Lemma 3.22. *Fix $p > 0$, $\alpha < 2$, and let $\{X_t : t \geq 0\}$ be a Lévy process with $X_1 \sim TS^p_\alpha(R, b)$. If $R \neq 0$, then the Blumenthal-Getoor index of this process is*

$$\beta = \alpha \vee r, \tag{3.25}$$

where

$$r = \inf\left\{\gamma > 0 : \int_{|x|\leq 1} |x|^\gamma R(\mathrm{d}x) < \infty\right\}.$$

This follows immediately from the following.

Lemma 3.23. *Fix $\alpha < 2$, $p > 0$, let R be the Rosiński measure of a p-tempered α-stable distribution, and let M be the corresponding Lévy measure. If $R \neq 0$, then for any $q \in (-\infty, 2)$*

$$\int_{|x|\leq 1} |x|^q M(\mathrm{d}x) < \infty \iff \alpha < q \text{ and } \int_{|x|\leq 1} |x|^q R(\mathrm{d}x) < \infty.$$

Proof. First assume that $\int_{|x|\leq 1} |x|^q M(\mathrm{d}x) < \infty$ and choose $r > 0$ such that $R(|x| \leq r) > 0$. We have

$$\begin{aligned}
\infty &> \int_{|x|\leq 1} |x|^q M(\mathrm{d}x) \\
&\geq \int_{|x|\leq r} |x|^q \int_0^{|x|^{-1}} t^{q-\alpha-1} e^{-t^p} \mathrm{d}t R(\mathrm{d}x) \\
&\geq e^{-r^{-p}} \int_{|x|\leq r} |x|^q R(\mathrm{d}x) \int_0^{r^{-1}} t^{q-\alpha-1} \mathrm{d}t,
\end{aligned}$$

which implies that $\alpha < q$ and $\int_{|x|\leq 1} |x|^q R(\mathrm{d}x) < \infty$. Now assume that $\alpha < q$ and $\int_{|x|\leq 1} |x|^q R(\mathrm{d}x) < \infty$. We have

$$\begin{aligned}
\int_{|x|\leq 1} |x|^q M(\mathrm{d}x) &= \int_{\mathbb{R}^d} |x|^q \int_0^{|x|^{-1}} t^{q-\alpha-1} e^{-t^p} \mathrm{d}t R(\mathrm{d}x) \\
&\leq \int_{|x|\leq 1} |x|^q R(\mathrm{d}x) \int_0^\infty t^{q-\alpha-1} e^{-t^p} \mathrm{d}t + \int_{|x|>1} |x|^q \int_0^{|x|^{-1}} t^{q-\alpha-1} \mathrm{d}t R(\mathrm{d}x) \\
&\leq \int_{|x|\leq 1} |x|^q R(\mathrm{d}x) \int_0^\infty t^{q-\alpha-1} e^{-t^p} \mathrm{d}t + (q-\alpha)^{-1} \int_{|x|>1} |x|^\alpha R(\mathrm{d}x),
\end{aligned}$$

which is finite. □

Combining Lemma 3.22 with (3.15) tells us that the Blumenthal-Getoor index of a proper p-tempered α-stable Lévy processes with $\alpha \in (0,2)$ is α. It may be interesting to note that α is also the Blumenthal-Getoor index of any α-stable Lévy process, see, e.g., [12]. We now discuss several properties that are characterized by this index.

Let $X = \{X_t : t \geq 0\}$ be a Lévy process with $X_1 \sim TS^p_\alpha(R,b)$ and let β be given by (3.25). From [12] it follows that, with probability 1,

$$\limsup_{t\to 0} t^{-1/\gamma}|X_t| = \begin{cases} \infty & \text{if } \gamma < \beta \\ 0 & \text{if } \gamma > \beta \end{cases}.$$

Now, fix $0 \leq a < b < \infty$, $\gamma > 0$, and define

$$V_\gamma(X;a,b) = \sup \sum_{j=1}^{n} |X_{t_i} - X_{t_{i-1}}|^\gamma,$$

where the supremum is taken over all finite partitions $a = t_0 < t_1 < \cdots < t_{n-1} < t_n = b$ of the interval $[a,b]$. This is called the **γ-variation** of X. From [12] and [56] it follows that for any $0 \leq a < b < \infty$ with probability 1

$$V_\gamma(X;a,b) \begin{cases} = \infty & \text{if } \gamma < \beta \\ < \infty & \text{if } \gamma > \beta \end{cases}. \tag{3.26}$$

Finiteness of γ-variation gives useful results about how one can define stochastic integrals with respect to these processes. It is well known that if a process has finite 1-variation, then one can define a Stieltjes integral with respect to it. When the 1-variation is infinite, under certain assumptions about the finiteness of γ-variation for some $\gamma > 0$, one can define generalizations of Stieltjes integrals, see [22] for details.

We sometimes refer to 1-variation as simply **variation**. Thus (3.26) and Lemma 3.22 imply that a p-tempered α-stable Lévy process has finite variation if and only if $\alpha < 1$ and $\int_{|x|\leq 1}|x|R(\mathrm{d}x) < \infty$. In particular, in light of (3.15), all proper p-tempered α-stable Lévy processes with $\alpha < 1$ have finite variation. We now turn to a related concept.

A one-dimensional Lévy process, which is nondecreasing almost surely is called a **subordinator**. Such a process necessarily has finite variation. Further, by combining the above discussion with Theorems 21.5 and 21.9 in [69] we can fully characterize when a p-tempered α-stable Lévy process is a subordinator.

Proposition 3.24. *Let $\{X_t : t \geq 0\}$ be a one-dimensional Lévy process with $X_1 \sim TS^p_\alpha(R,b)$ with $R \neq 0$. The process is a subordinator if and only if $\alpha < 1$, $R((-\infty,0)) = 0$, $\int_{(0,1)} xR(\mathrm{d}x) < \infty$, and $b \geq \int_{(0,\infty)}\int_0^\infty \frac{x}{1+t^2x^2}t^{-\alpha}e^{-t^p}\mathrm{d}tR(\mathrm{d}x)$.*

Remark 3.6. A Lévy process is a subordinator if and only if the distribution of X_t has its support contained in $[0,\infty)$ for every t. Further, if $\{X_t : t \geq 0\}$ is a subordinator with $X_1 \sim TS^p_\alpha(R,b)$ and $R \neq 0$ then, by Theorem 24.10 in [69], the support of the distribution of X_t is given by $[t\zeta,\infty)$, where $\zeta = b - \int_{(0,\infty)}\int_0^\infty \frac{x}{1+t^2x^2}t^{-\alpha}e^{-t^p}\mathrm{d}tR(\mathrm{d}x)$.

We conclude this section by discussing when the distribution of a proper p-tempered α-stable Lévy process (with $\alpha \in (0,2)$) is absolutely continuous with respect to the distribution of the α-stable Lévy process that is being tempered. Our presentation follows [66] closely. Let $\Omega = D([0,\infty),\mathbb{R}^d)$ be the space of mappings $\omega(\cdot)$ from $[0,\infty)$ into $\mathbb{R}^d$ that are right-continuous with left limits. Let $X = \{X_t : t \geq 0\}$ be the collection of functions from Ω into $\mathbb{R}^d$ with $X_t(\omega) = \omega(t)$. Assume that Ω be equipped with the σ-algebra $\mathscr{F} = \sigma(X_s : s \geq 0)$ and the right-continuous natural filtration $(\mathscr{F}_t)_{t\geq 0}$ where $\mathscr{F}_t = \bigcap_{s>t} \sigma(X_u : u \leq s)$. In this case X is called **the canonical process**. The distribution of this process is completely determined by a probability measure P on $(\Omega, \mathscr{F})$. Let $P_{|\mathscr{F}_t}$ denote the restriction of P to the σ-algebra $\mathscr{F}_t$.

Theorem 3.25. *Fix $\alpha \in (0,2)$ and $p > 0$. In the above setting, consider two probability measures P_0 and P on $(\Omega, \mathscr{F})$ and let $X = \{X_t : t \geq 0\}$ be the canonical process. Assume that, under P, X is a Lévy process with $X_1 \sim TS^p_\alpha(R,b)$, where R satisfies (3.15).[3] Derive σ from R by (3.17) and let $q(u,r)$ be as in Proposition 3.6. If, under P_0, X is a Lévy process with $X_1 \sim S_\alpha(a,\sigma)$, then*

1. *$P_{0|\mathscr{F}_t}$ and $P_{|\mathscr{F}_t}$ are mutually absolutely continuous for every $t > 0$ if and only if*

$$\int_{\mathbb{S}^{d-1}} \int_0^1 [1 - q(r^p, u)]^2 r^{-\alpha-1} \mathrm{d}r \sigma(\mathrm{d}u) < \infty \tag{3.27}$$

and

$$b - a = \int_{\mathbb{R}^d} \int_0^\infty \frac{x}{1+|x|^2 t^2} t^{-\alpha} (e^{-t^p} - 1) \mathrm{d}t R(\mathrm{d}x). \tag{3.28}$$

2. *If $P_{0|\mathscr{F}_t}$ and $P_{|\mathscr{F}_t}$ are not mutually absolutely continuous for some $t > 0$, then they are singular for all $t > 0$.*
3. *If (3.27) and (3.28) hold, then for every $t > 0$*

$$\frac{\mathrm{d}P_{|\mathscr{F}_t}}{\mathrm{d}P_{0|\mathscr{F}_t}} = e^{U_t}, \; P_0 \; a.s.$$

where

$$U_t = \lim_{\epsilon \downarrow 0} \left\{ \sum_{\{s \in (0,t] : |\Delta X_s| > \epsilon\}} \log q\left(|\Delta X_s|^p, \frac{\Delta X_s}{|\Delta X_s|} \right) + t \int_{\mathbb{S}^{d-1}} \int_\epsilon^\infty [1 - q(r^p, u)] r^{-\alpha-1} \mathrm{d}r \sigma(\mathrm{d}u) \right\}.$$

[3] This implies that X_1 has a proper p-tempered α-stable distribution.

and the convergence is uniform in t on any bounded interval, P_0 a.s. Further, $\{U_t : t \geq 0\}$ is a one-dimensional Lévy process defined on the probability space $(\Omega, \mathscr{F}, P_0)$. It satisfies $U_1 \sim ID(0, M_U, b_U)$, where

$$M_U(A) = \int_{\mathbb{S}^{d-1}} \int_0^\infty 1_{A \setminus \{0\}}(\log[q(r^p, u)]) r^{-\alpha-1} \mathrm{d}r\sigma(\mathrm{d}u), \qquad A \in \mathfrak{B}(\mathbb{R})$$

and

$$b_U = -\int_{-\infty}^0 \left(e^y - 1 - \frac{y}{1+|y|^2} \right) M_U(\mathrm{d}y).$$

Note that Proposition 3.6 implies that $q(r^p, u) \in (0, 1]$, and hence that M_U satisfies $M_U([0, \infty)) = 0$.

Proof. By Remark 3.5 all proper p-tempered α-stable distributions with $\alpha \in (0, 2)$ belong to the class of generalized tempered α-stable distributions. For these, analogues of Parts 1 and 2 are given in Theorem 4.1 of [66]. In [66] the analogue of (3.27) is actually

$$\int_{\mathbb{S}^{d-1}} \int_0^1 (1 - [q(r^p, u)]^{1/2})^2 r^{-\alpha-1} \mathrm{d}r\sigma(\mathrm{d}u) < \infty.$$

As observed in [65], this is equivalent to (3.27) since for any $x \in [0, 1]$

$$.25(1-x)^2 \leq (1 - \sqrt{x})^2 \leq (1-x)^2,$$

and $q(r^p, u) \in [0, 1]$ for all r and u. For any Lévy process, an analogue of Part 3 is given in Theorem 33.2 of [69]. To specialize it to our situation we just need to apply (3.2). □

Under additional conditions, representations of the process U_t in terms of certain extensions of γ-variation can be given, see [24]. As pointed out in [65] condition (3.27) fails when the function $q(r^p, u)$ decreases too quickly near zero. In other words when there is too much tempering near zero. This is illustrated by the following.

Corollary 3.26. *Fix $\alpha \in (0, 2)$ and $p > 0$. Let P_0, P, and $\{X_t : t \geq 0\}$ be as in Theorem 3.25. If $p \leq \alpha/2$, then $P_{0|\mathscr{F}_t}$ and $P_{|\mathscr{F}_t}$ are mutually singular for all $t > 0$.*

Proof. By Remark 3.4 we can write $q(r^p, u) = \int_{(0,\infty)} e^{-r^p s} Q_u(\mathrm{d}s)$ for some measurable family of probability measures $\{Q_u\}_{u \in \mathbb{S}^{d-1}}$. Since $1 - e^{-x} \geq \frac{x}{1+x}$ for any $x \geq 0$ (see, e.g., 4.2.32 in [2]) it follows that

$$\int_{\mathbb{S}^{d-1}} \int_0^1 [1 - q(r^p, u)]^2 r^{-\alpha-1} \mathrm{d}r\sigma(\mathrm{d}u)$$
$$= \int_{\mathbb{S}^{d-1}} \int_0^1 \left[\int_{(0,\infty)} (1 - e^{-r^p s}) Q_u(\mathrm{d}s) \right]^2 r^{-\alpha-1} \mathrm{d}r\sigma(\mathrm{d}u)$$

$$\geq \int_{\mathbb{S}^{d-1}} \int_0^1 \left[\int_{(0,\infty)} \frac{r^p s}{1 + r^p s} Q_u(ds) \right]^2 r^{-\alpha-1} dr \sigma(du)$$

$$\geq \int_{\mathbb{S}^{d-1}} \left[\int_{(0,\infty)} \frac{s}{1+s} Q_u(ds) \right]^2 \sigma(du) \int_0^1 r^{2p-\alpha-1} dr,$$

which equals infinity when $p \leq \alpha/2$. From here the result follows by Part 2 of Theorem 3.25. □

3.5 Exponential Moments When $p = 1$

A representation for the exponential moments of p-tempered α-stable distributions is given by (3.24). In this section we derive significantly simpler formulas for the case[4] where $p = 1$. Throughout this section we use the principle branch of the complex logarithm, i.e. we make a cut along the negative real axis. This implies that for $z \in \mathbb{C}$ with $\Re z > 0$ we have $\log(z) = \log|z| + i \arctan(\Im z / \Re z)$, where arctan refers to the branch of the arctangent whose image is $\left(-\frac{\pi}{2}, \frac{\pi}{2}\right)$. We begin with a lemma.

Lemma 3.27. *Fix $\alpha < 2$, $p = 1$, and $\mu = TS_\alpha^1(R, b)$. Let $X \sim \mu$, let S be the support[5] of R, and fix $z \in \mathbb{C}^d$. When $\alpha \in (0, 2)$ we have*

$$\mathrm{E}\left|e^{\langle z, X\rangle}\right| < \infty \tag{3.29}$$

if and only if $\sup_{x \in S} \Re\langle z, x\rangle \leq 1$. When $\alpha \leq 0$ a sufficient[6] condition for (3.29) is $\sup_{x \in S} \Re\langle z, x\rangle < 1$.

Proof. We will need the following fact from 6.1.1 in [2]. When $\alpha < 0$ and $w \in \mathbb{C}$ with $\Re w > 0$ we have

$$\int_0^\infty e^{-wt} t^{-\alpha-1} dt = w^\alpha \Gamma(-\alpha). \tag{3.30}$$

By Theorem 25.17 in [69] (3.29) is equivalent to $\int_{|x|>1} e^{\langle c, x\rangle} M(dx) < \infty$ where $c = \Re z$ and M is the Lévy measure of μ. When $c = 0$ this always holds so assume that $c \neq 0$. By (3.12) we have

---

[4] The only other case where reasonable representations are known is when $p = 2$ and $\alpha \in (0, 2)$. In this case [9] gives formulas in terms of confluent hypergeometric functions.

[5] This means that S is the smallest closed subset of $\mathbb{R}^d$ with $R(S^c) = 0$.

[6] In light of Theorem 3.17, it is clear that this is not a necessary condition when $\alpha \leq 0$.

$$
\begin{aligned}
\int_{|x|>1} e^{\langle c,x\rangle} M(\mathrm{d}x) &= \int_S \int_{|x|^{-1}}^{\infty} e^{\langle c,x\rangle t} e^{-t} t^{-1-\alpha} \mathrm{d}t R(\mathrm{d}x) \\
&= \int_{S\cap[|x|\le 1/(2|c|)]} \int_{|x|^{-1}}^{\infty} e^{\langle c,x\rangle t} e^{-t} t^{-1-\alpha} \mathrm{d}t R(\mathrm{d}x) \\
&\quad \int_{S\cap[|x|>1/(2|c|])} \int_{|x|^{-1}}^{\infty} e^{\langle c,x\rangle t} e^{-t} t^{-1-\alpha} \mathrm{d}t R(\mathrm{d}x) \\
&=: I_1(\alpha) + I_2(\alpha).
\end{aligned}
$$

Let $K := \sup_{t\ge 2|c|} e^{-t/2} t^{2-\alpha}$ and note that for every $\alpha < 2$ we have

$$
\begin{aligned}
I_1(\alpha) &\le \int_{|x|\le 1/(2|c|)} \int_{|x|^{-1}}^{\infty} e^{t/2} e^{-t} t^{-1-\alpha} \mathrm{d}t R(\mathrm{d}x) \\
&\le K \int_{|x|\le 1/(2|c|)} \int_{|x|^{-1}}^{\infty} t^{-3} \mathrm{d}t R(\mathrm{d}x) = .5K \int_{|x|\le 1/(2|c|)} |x|^2 R(\mathrm{d}x) < \infty.
\end{aligned}
$$

Thus, finiteness of the exponential moment is determined by $I_2(\alpha)$. Define $\theta = \sup_{x\in S}\langle c, x\rangle$. We begin with the case $\alpha \in (0, 2)$. If $\theta \le 1$, then

$$
\begin{aligned}
I_2(\alpha) &\le \int_{S\cap[|x|>1/(2|c|])} \int_{|x|^{-1}}^{\infty} e^{\theta t} e^{-t} t^{-1-\alpha} \mathrm{d}t R(\mathrm{d}x) \\
&\le \int_{|x|>1/(2|c|)} \int_{|x|^{-1}}^{\infty} t^{-1-\alpha} \mathrm{d}t R(\mathrm{d}x) = \alpha^{-1} \int_{|x|>1/(2|c|)} |x|^{\alpha} \mathrm{d}t R(\mathrm{d}x) < \infty.
\end{aligned}
$$

On the other hand, if $\theta > 1$, then there is an $\epsilon > 0$ and a Borel set $S_\epsilon \subset S \cap [|x| > 1/(2|c|)]$ with $R(S_\epsilon) > 0$ such that for every $x \in S_\epsilon$ we have $\langle x, c\rangle \ge 1 + \epsilon$. This implies that

$$
\begin{aligned}
I_2(\alpha) &\ge \int_{S_\epsilon} \int_{|x|^{-1}}^{\infty} e^{\langle c,x\rangle t} e^{-t} t^{-1-\alpha} \mathrm{d}t R(\mathrm{d}x) \\
&\ge \int_{S_\epsilon} \int_{2|c|}^{\infty} e^{(1+\epsilon)t} e^{-t} t^{-1-\alpha} \mathrm{d}t R(\mathrm{d}x) \\
&= R(S_\epsilon) \int_{2|c|}^{\infty} e^{\epsilon t} t^{-1-\alpha} \mathrm{d}t = \infty.
\end{aligned}
$$

Now assume that $\alpha \le 0$ and $\theta < 1$. For $\alpha < 0$ we can use (3.30) to get

$$
\begin{aligned}
I_2(\alpha) &\le \int_{|x|>1/(2|c|)} \int_0^{\infty} e^{-t(1-\theta)} t^{-1-\alpha} \mathrm{d}t R(\mathrm{d}x) \\
&= (1-\theta)^{-|\alpha|} \Gamma(|\alpha|) \int_{|x|>1/(2|c|)} R(\mathrm{d}x) < \infty,
\end{aligned}
$$

and for $\alpha = 0$ we get

$$I_2(0) \le \int_{|x|>1/(2|c|)} \int_{|x|^{-1}}^{2|c|} t^{-1}\mathrm{d}tR(\mathrm{d}x) + R\left(|x| > \frac{1}{2|c|}\right) \int_{2|c|}^{\infty} e^{-t(1-\theta)}t^{-1}\mathrm{d}t$$
$$= \int_{|x|>1/(2|c|)} \log\left(2|c||x|\right) R(\mathrm{d}x) + R\left(|x| > \frac{1}{2|c|}\right) \int_{2|c|}^{\infty} e^{-t(1-\theta)}t^{-1}\mathrm{d}t < \infty,$$

which completes the proof. □

We now give the main result of this section.

Theorem 3.28. *Fix $\alpha < 2$, $p = 1$, $\mu = TS_\alpha^1(R, b)$, and let $X \sim \mu$. Let S be the support of R and fix $z \in \mathbb{C}^d$ such that either a) $\sup_{x\in S} \Re\langle z, x\rangle < 1$ or b) $\Im z = 0$, $\sup_{x\in S} \Re\langle z, x\rangle \le 1$, and $\alpha \in (0, 2)$. In both cases (3.29) holds and we have:*

1. *If $\int_{\mathbb{R}^d} |x|\mu(\mathrm{d}x) < \infty$, then*

$$\mathrm{E}e^{\langle z,X\rangle} = \exp\left\{\int_{\mathbb{R}^d} \psi_\alpha(\langle z, x\rangle)R(\mathrm{d}x) + \langle z, b_1\rangle\right\}, \tag{3.31}$$

where

$$b_1 = b + \int_{\mathbb{R}^d}\int_0^\infty x\frac{|x|^2}{1+|x|^2t^2}t^{2-\alpha}e^{-t}\mathrm{d}tR(\mathrm{d}x) \tag{3.32}$$

and

$$\psi_\alpha(s) = \begin{cases} \Gamma(-\alpha)[(1-s)^\alpha - 1 + \alpha s] & \alpha \ne 0, 1 \\ -\log(1-s) - s & \alpha = 0 \\ (1-s)\log(1-s) + s & \alpha = 1 \end{cases}. \tag{3.33}$$

In particular this holds when $1 < \alpha < 2$, or

$$\alpha = 1 \quad \text{and} \quad \int_{|x|>1} |x| \log|x| R(\mathrm{d}x) < \infty,$$

or

$$\alpha < 1 \quad \text{and} \quad \int_{\mathbb{R}^d} |x|R(\mathrm{d}x) < \infty.$$

2. *If $\alpha < 1$ and $\int_{|x|\le 1} |x|R(\mathrm{d}x) < \infty$, then*

$$\mathrm{E}e^{\langle z,X\rangle} = \exp\left\{\int_{\mathbb{R}^d} \psi_\alpha^0(\langle z, x\rangle)R(\mathrm{d}x) + \langle z, b_0\rangle\right\}, \tag{3.34}$$

where

$$b_0 = b - \int_{\mathbb{R}^d}\int_0^\infty \frac{x}{1+|x|^2t^2} t^{-\alpha} e^{-t} \mathrm{d}t R(\mathrm{d}x) \tag{3.35}$$

and

$$\psi_\alpha^0(s) = \begin{cases} \Gamma(-\alpha)[(1-s)^\alpha - 1] & \alpha \neq 0 \\ -\log(1-s) & \alpha = 0 \end{cases}. \tag{3.36}$$

In particular, this holds if μ is a proper TS_α^1 distribution with $\alpha < 1$.

In the above we take $\Psi_1(1) = 1$, which is the limiting value of the function $\Psi_\alpha(s)$ in both s and α. A simple way to ensure that the assumption of Theorem 3.28 holds is as follows. Fix $\theta > 0$. If $R(|x| > \theta^{-1}) = 0$, then for any $z \in \mathbb{C}^d$ with $|\Re z| < \theta$ we have $\sup_{x\in S}\langle \Re z, x\rangle \le \sup_{x\in S} |\langle \Re z, x\rangle| \le \sup_{x\in S} |\Re z||x| < \theta/\theta = 1$. The vectors b_1 and b_0 given above have the following interpretations. When $\int_{\mathbb{R}^d} |x|\mu(\mathrm{d}x) < \infty$ we have $b_1 = \int_{\mathbb{R}^d} x\mu(\mathrm{d}x)$, and when $\alpha < 1$ and $\int_{|x|\le 1} |x|R(\mathrm{d}x) < \infty$ the vector b_0 is the drift.

Proof. Our proof will use the following. If $t \in (0,1)$, $s \in \mathbb{C}$, and $\alpha \le 1$, then

$$\begin{aligned} |(e^{st} - 1 - st)t^{-\alpha-1}e^{-t}| &\le \sum_{n=2}^\infty \frac{|st|^n}{n!} t^{-\alpha-1} e^{-t} \\ &= t^{1-\alpha} e^{-t} |s|^2 \sum_{n=2}^\infty \frac{|st|^{n-2}}{n(n-1)(n-2)!} \\ &\le e^{-t}|s|^2 \sum_{n=2}^\infty \frac{|st|^{n-2}}{(n-2)!} \\ &= e^{t(|s|-1)}|s|^2. \end{aligned} \tag{3.37}$$

Lemma 3.27 implies that we can use (3.24) to get a representation for the exponential moment. We begin with the case $\int_{\mathbb{R}^d} |x|\mu(\mathrm{d}x) < \infty$. In this case b_1 is definable as a vector in $\mathbb{R}^d$ and from (3.24) it follows that

$$\begin{aligned} \mathrm{E}e^{\langle z, X\rangle} &= \exp\left\{\int_{\mathbb{R}^d}\int_0^\infty \left(e^{\langle x,z\rangle t} - 1 - \langle x, z\rangle t\right) t^{-1-\alpha} e^{-t} \mathrm{d}t R(\mathrm{d}x) + \langle z, b_1\rangle\right\} \\ &= \exp\left\{\int_S\int_0^\infty \left(e^{\langle x,z\rangle t} - 1 - \langle x, z\rangle t\right) t^{-1-\alpha} e^{-t} \mathrm{d}t R(\mathrm{d}x) + \langle z, b_1\rangle\right\}. \end{aligned}$$

Fix $x \in S$. For simplicity of notation let $s = \langle x, z\rangle$ and note that, by assumption, when $\alpha \le 0$ we have $\Re s < 1$ and $\Re(1-s) > 0$ and when $\alpha \in (0,2)$ we have $\Re s \le 1$ and $\Re(1-s) \ge 0$. When $\alpha < 0$ we can use (3.30) to get

$$\int_0^\infty (e^{st} - 1 - st)e^{-t}t^{-\alpha-1}\mathrm{d}t = \int_0^\infty (e^{-(1-s)t} - e^{-t} - se^{-t}t)t^{-\alpha-1}\mathrm{d}t$$
$$= \Gamma(-\alpha)[(1-s)^\alpha - 1 + \alpha s].$$

When $\alpha = 0$ we can use l'Hôpital's rule[7] to get

$$\begin{aligned}\int_0^\infty (e^{st} - 1 - st)e^{-t}t^{-1}\mathrm{d}t &= \int_0^\infty \lim_{\alpha\uparrow 0}(e^{st} - 1 - st)e^{-t}t^{-\alpha-1}\mathrm{d}t \\ &= \lim_{\alpha\uparrow 0}\int_0^\infty (e^{st} - 1 - st)e^{-t}t^{-\alpha-1}\mathrm{d}t \\ &= \lim_{\alpha\uparrow 0}\Gamma(-\alpha)[(1-s)^\alpha - 1 + \alpha s] \\ &= \lim_{\alpha\uparrow 0}\frac{\Gamma(1-\alpha)[(1-s)^\alpha - 1 + \alpha s]}{-\alpha} \\ &= \lim_{\alpha\uparrow 0}\frac{(1-s)^\alpha - 1 + \alpha s}{-\alpha} \\ &= -\log(1-s) - s,\end{aligned}$$

where we can interchange limit and integral using dominated convergence. Specifically, for $\alpha \in (-1, 0)$ if $t \in (0, 1)$, then (3.37) gives a bound that is integrable on $(0, 1)$ and if $t \geq 1$, then

$$|(e^{st} - 1 - st)t^{-\alpha-1}e^{-t}| \leq e^{-t(1-\Re s)} + (1 + |s|t)e^{-t},$$

which is integrable on $[1, \infty)$ since $\Re s < 1$.

Now assume that $\alpha \in (0, 1)$. For any $v, w \in \mathbb{C}$ with w satisfying $\Re w > 0$ and v satisfying either $\Re v > 0$ or $v = 0$ integration by parts and (3.30) give

$$\int_0^\infty \left(e^{-vt} - e^{-wt}\right)t^{-1-\alpha}\mathrm{d}t = \Gamma(-\alpha)(v^\alpha - w^\alpha), \tag{3.38}$$

which implies

$$\int_0^\infty (e^{st} - 1 - st)e^{-t}t^{-\alpha-1}\mathrm{d}t = \int_0^\infty (e^{-(1-s)t} - e^{-t})t^{-\alpha-1}\mathrm{d}t - s\int_0^\infty e^{-t}t^{(1-\alpha)-1}\mathrm{d}t$$
$$= \Gamma(-\alpha)[(1-s)^\alpha - 1 + s\alpha].$$

[7] We can use l'Hôpital's rule because the denominator is real. However, in general, l'Hôpital's rule may fail for complex valued functions of real numbers, see [18].

Now assume that $\alpha \in (1,2)$. For any $v, w \in \mathbb{C}$ with w satisfying $\Re w > 0$ and v satisfying either $\Re v > 0$ or $v = 0$ integration by parts and (3.38) give

$$\int_0^\infty \left(e^{-vt} - e^{-wt} + (v-w)t\right) t^{-1-\alpha} \mathrm{d}t = \Gamma(-\alpha)(v^\alpha - w^\alpha),$$

which implies

$$\begin{aligned}\int_0^\infty (e^{st} - 1 - st)e^{-t}t^{-\alpha-1}\mathrm{d}t &= \int_0^\infty (e^{-(1-s)t} - e^{-t} - st)t^{-\alpha-1}\mathrm{d}t \\ &\quad + s\int_0^\infty (1-e^{-t})t^{(1-\alpha)-1}\mathrm{d}t \\ &= \Gamma(-\alpha)[(1-s)^\alpha - 1 + s\alpha].\end{aligned}$$

Now consider the case $\alpha = 1$. By l'Hôpital's rule

$$\begin{aligned}\int_0^\infty (e^{st} - 1 - st)e^{-t}t^{-2}\mathrm{d}t &= \int_0^\infty \lim_{\alpha\uparrow 1}(e^{st} - 1 - st)e^{-t}t^{-\alpha-1}\mathrm{d}t \\ &= \lim_{\alpha\uparrow 1}\int_0^\infty (e^{st} - 1 - st)e^{-t}t^{-\alpha-1}\mathrm{d}t \\ &= \lim_{\alpha\uparrow 1}\Gamma(-\alpha)[(1-s)^\alpha - 1 + s\alpha] \\ &= \lim_{\alpha\uparrow 1}\frac{\Gamma(2-\alpha)}{(\alpha-1)\alpha}[(1-s)^\alpha - 1 + s\alpha] \\ &= (1-s)\log(1-s) + s,\end{aligned}$$

where the second line follows by dominated convergence. Specifically, for $\alpha \in (.5, 1)$ if $t \in (0,1)$, then (3.37) gives a bound that is integrable on $(0,1)$, and for $t \geq 1$ we have

$$|(e^{st} - 1 - st)t^{-\alpha-1}e^{-t}| \leq e^{-t(1-\Re s)}t^{-.5-1} + (1+|s|)e^{-t},$$

which is integrable on $[1,\infty)$ since $\Re s \leq 1$.

We now turn to the case when $\int_{|x|\leq 1}|x|R(\mathrm{d}x) < \infty$ and $\alpha < 1$. In this case b_0 is definable as a vector in $\mathbb{R}^d$ and (3.24) implies that

$$\mathrm{E}e^{\langle z,X\rangle} = \exp\left\{\int_{\mathbb{R}^d}\int_0^\infty \left(e^{\langle x,z\rangle t} - 1\right)t^{-1-\alpha}e^{-t}\mathrm{d}tR(\mathrm{d}x) + \langle z, b_0\rangle\right\}.$$

The fact that $\int_0^\infty \left(e^{\langle x,z\rangle t} - 1\right)t^{-1-\alpha}e^{-t}\mathrm{d}t$ has the required form can be shown in a similar way to the previous part. The conditions to guarantee $\int_{\mathbb{R}^d}|x|\mu(\mathrm{d}x) < \infty$ follow from Theorem 3.15, while the fact that $\int_{|x|\leq 1}|x|R(\mathrm{d}x) < \infty$ for all proper TS_α^1 distribution with $\alpha < 1$ follows by Theorem 3.3. □

Note that the assumption of Theorem 3.28 always holds when $\Re z = 0$. This gives the following representation for the characteristic function.

Corollary 3.29. *Fix $\alpha < 2$, $p = 1$, and let $\mu = TS^1_\alpha(R, b)$.*

1. *If $\int_{\mathbb{R}^d} |x|\mu(\mathrm{d}x) < \infty$, then*

$$\hat{\mu}(z) = \exp\left\{\int_{\mathbb{R}^d} \psi_\alpha(i\langle z, x\rangle)R(\mathrm{d}x) + i\langle z, b_1\rangle\right\}, \qquad z \in \mathbb{R}^d,$$

where b_1 is given by (3.32) and ψ_α is given by (3.33).

2. *If $\alpha < 1$ and $\int_{|x|\le 1} |x|R(\mathrm{d}x) < \infty$, then the characteristic function is given by*

$$\hat{\mu}(z) = \exp\left\{\int_{\mathbb{R}^d} \psi^0_\alpha(i\langle z, x\rangle)R(\mathrm{d}x) + i\langle z, b_0\rangle\right\}, \qquad z \in \mathbb{R}^d,$$

where b_0 is given by (3.35) and ψ^0_α is given by (3.36).

Now consider the case when $X \sim TS^1_\alpha(R, b)$ is a one-dimensional random variable with $R((-\infty, 0)) = 0$. In this case the support, S, of R satisfies $S \subset [0, \infty)$. Thus for all $z \in \mathbb{R}$ with $z \le 0$ we have $\sup_{x\in S}(zx) < 1$ and we can use Theorem 3.28 to get the following representation for the Laplace transform.

Corollary 3.30. *Fix $\alpha < 2$, $p = 1$, let $\mu = TS^1_\alpha(R, b)$ be a 1-tempered α-stable distribution on $\mathbb{R}$ with $R((-\infty, 0)) = 0$, and let $X \sim \mu$.*

1. *If $\mathrm{E}|X| < \infty$, then*

$$\mathrm{E}[e^{-zX}] = \exp\left\{\int_{(0,\infty)} \psi_\alpha(-zx)R(\mathrm{d}x) - zb_1\right\}, \quad z \ge 0,$$

where b_1 is given by (3.32) and ψ_α is given by (3.33).

2. *If $\alpha < 1$ and $\int_{|x|\le 1} |x|R(\mathrm{d}x) < \infty$, then*

$$\mathrm{E}[e^{-zX}] = \exp\left\{\int_{(0,\infty)} \psi^0_\alpha(-zx)R(\mathrm{d}x) - zb_0\right\}, \quad z \ge 0,$$

where b_0 is given by (3.35) and ψ^0_α is given by (3.36).

Chapter 4
Limit Theorems for Tempered Stable Distributions

In this chapter we discuss the weak limits of sequences of p-tempered α-stable distributions. It turns out that this class is not closed under weak convergence. To see this note that the class has elements with a finite variance (see Theorem 3.15) and is closed under shifting, scaling, and taking convolutions (see Proposition 3.5). Thus, by the central limit theorem, there are sequences in this class that converge weakly to Gaussian distributions. However, Gaussian distributions were explicitly excluded from the class. Further, as we will see, we also need to include α-stable distributions, which do not belong to this class by Proposition 3.12. Thus, to get closure under weak convergence we need to extend the class.

4.1 Extended Tempered Stable Distributions

In this section we define a class of distributions, which, as we will show, is the smallest class that contains TS^p_α and is closed under weak convergence. To do this we must allow for a Gaussian part and remove the assumption that (3.4) holds. We will see that removing this assumption is equivalent to allowing for an α-stable part.

Definition 4.1. Fix $\alpha < 2$ and $p > 0$. An infinitely divisible probability measure μ is called an **extended p-tempered α-stable distribution** if its Lévy measure is given by (3.2) where σ is a finite Borel measure on $\mathbb{S}^{d-1}$ and $q : (0, \infty) \times \mathbb{S}^{d-1} \mapsto (0, \infty)$ is a Borel function such that for all $u \in \mathbb{S}^{d-1}$ $q(\cdot, u)$ is completely monotone and (3.3) holds. We denote the class of extended p-tempered α-stable distributions by ETS^p_α.

Remark 4.1. When $\alpha \leq 0$ (3.4) is necessary for (3.3) to hold. Thus, whenever $\alpha \leq 0$ an ETS^p_α distribution is just a TS^p_α distribution with a Gaussian part.

Remark 4.2. For $\alpha < 2$ and $p > 0$, the class of distributions in TS^p_α but allowing for a Gaussian part was introduced in [51] under the name $J_{\alpha,p}$. By Remark 4.1 when $\alpha \leq 0$ we have $J_{\alpha,p} = ETS^p_\alpha$. However, when $\alpha \in (0, 2)$ we have $J_{\alpha,p} \subsetneq ETS^p_\alpha$.

M. Grabchak, *Tempered Stable Distributions*, SpringerBriefs in Mathematics, DOI 10.1007/978-3-319-24927-8_4

Remark 4.3. From (3.5) it follows that the sum of completely monotone functions is completely monotone, which implies that the class ETS^p_α is closed under taking convolutions.

By Bernstein's Theorem (see, e.g., Theorem 1a in Section XIII.4 of [23] or Remark 3.2 in [6]) the complete monotonicity of $q(\cdot, u)$ implies that there is a measurable family $\{Q_u\}_{u\in\mathbb{S}^{d-1}}$ of Borel measures on $[0,\infty)$ such that

$$q(r,u) = \int_{[0,\infty)} e^{-rs} Q_u(\mathrm{d}s). \tag{4.1}$$

Note that unlike (3.6) we now allow Q_u to have a point mass at 0. By the Dominated Convergence Theorem, it follows that

$$\lim_{r\to\infty} q(r,u) = Q_u(\{0\}). \tag{4.2}$$

Thus, by Remark 4.1, when $\alpha \le 0$ we have $Q_u(\{0\}) = 0$ for every $u \in \mathbb{S}^{d-1}$. Letting

$$q_1(r,u) = \int_{(0,\infty)} e^{-rs} Q_u(\mathrm{d}s) \tag{4.3}$$

gives

$$q(r^p,u) = q_1(r^p,u) + Q_u(\{0\}). \tag{4.4}$$

The Lévy measure of a distribution in ETS^p_α can now be written as

$$\begin{aligned} M(B) &= \int_{\mathbb{S}^{d-1}} \int_0^\infty 1_B(ru) q_1(r^p,u) r^{-\alpha-1} \mathrm{d}r \sigma(\mathrm{d}u) \\ &\quad + \int_{\mathbb{S}^{d-1}} \int_0^\infty 1_B(ru) r^{-\alpha-1} \mathrm{d}r Q_u(\{0\}) \sigma(\mathrm{d}u), \quad B \in \mathfrak{B}(\mathbb{R}^d). \end{aligned} \tag{4.5}$$

Note that M is the sum of the Lévy measure of a p-tempered α-stable distribution and (when $\alpha \in (0,2)$) an α-stable distribution with spectral measure $Q_u(\{0\})\sigma(\mathrm{d}u)$. If R is the Rosiński measure of the p-tempered α-stable part, then

$$\begin{aligned} M(B) &= \int_{\mathbb{R}^d} \int_0^\infty 1_B(rx) r^{-1-\alpha} e^{-r^p} \mathrm{d}r R(\mathrm{d}x) \\ &\quad + \int_{\mathbb{S}^{d-1}} \int_0^\infty 1_B(ru) r^{-1-\alpha} \mathrm{d}r Q_u(\{0\}) \sigma(\mathrm{d}u), \quad B \in \mathfrak{B}(\mathbb{R}^d). \end{aligned} \tag{4.6}$$

Remark 4.4. A distribution is an element of ETS^p_α if and only if it can be written as the convolution of a Gaussian distribution, an element of TS^p_α, and (when $\alpha \in (0,2)$) an α-stable distribution.

Note that M is defined in terms of two measures $R(\mathrm{d}x)$ and $Q_u(\{0\})\sigma(\mathrm{d}u)$. To make it easier to work with, we would like to combine these into one measure. Since R is already defined on all of $\mathbb{R}^d$ and $Q_u(\{0\})\sigma(\mathrm{d}u)$ requires a sphere, it makes sense to put it on a sphere at infinity. To do this we need to define an appropriate compactification of $\mathbb{R}^d$.

4.2 Interlude: A Compactification of $\mathbb{R}^d$

In this section we develop a compactification of $\mathbb{R}^d$ with a sphere at infinity and discuss vague convergence of Radon measures on this space. This will be fundamental for deriving weak limit theorems for distributions in the class ETS_α^p.

4.2.1 *Definitions*

Let $\mathbb{R}_0^d = \mathbb{R}^d \setminus \{0\}$ and note that for $x \in \mathbb{R}_0^d$ we have $x = |x|\frac{x}{|x|}$. Thus we can uniquely identify every element of $\mathbb{R}_0^d$ with an element of $(0,\infty) \times \mathbb{S}^{d-1}$. Let $\bar{\mathbb{R}}_0^d = (0,\infty] \times \mathbb{S}^{d-1}$ and $\bar{\mathbb{R}}^d = \bar{\mathbb{R}}_0^d \cup \{0\}$. For simplicity of notation define $\mathbb{I}^{d-1} = \{\infty\} \times \mathbb{S}^{d-1}$. For $u \in \mathbb{S}^{d-1}$ we write $\infty u = \{\infty\} \times \{u\}$ and for $D \subset \mathbb{S}^{d-1}$ we write $\infty D = \{\infty\} \times D$. We introduce the functions $\xi : \bar{\mathbb{R}}^d \mapsto \mathbb{S}^{d-1} \cup \{0\}$ and $\vartheta : \bar{\mathbb{R}}^d \mapsto [0,\infty]$ as follows. Let $\xi(0) = \vartheta(0) = 0$. If $x \in \bar{\mathbb{R}}_0^d$, then $x = \{r\} \times \{u\}$ and we define $\xi(x) = u$ and $\vartheta(x) = r$. For simplicity, we sometimes write $|x| := \vartheta(x)$. When $x \in \mathbb{I}^{d-1}$ we have $|x| = \infty$ and we adopt the convention $|x|^{-1} = 1/|x| = 0$. Let $\stackrel{\bar{\mathbb{R}}_+}{\to}$ and $\stackrel{\mathbb{R}^d}{\to}$ denote, respectively, the usual convergence on $[0,\infty]$ and on $\mathbb{R}^d$. If $x, x_1, x_2, \cdots \in \bar{\mathbb{R}}_0^d$, we write $x_n \to x$ or $\lim_{n\to\infty} x_n = x$ when $\vartheta(x_n) \stackrel{\bar{\mathbb{R}}_+}{\to} \vartheta(x)$ and $\xi(x_n) \stackrel{\mathbb{R}^d}{\to} \xi(x)$.

Let τ_0 be the topology induced by this definition of convergence (i.e., let τ_0 be the class of subsets of $\bar{\mathbb{R}}_0^d$ such that $A \in \tau_0$ if and only if for any $x \in A$ and any $x_1, x_2, \cdots \in \bar{\mathbb{R}}_0^d$ with $x_n \to x$ there is an N such that for all $n \ge N$, $x_n \in A$). In this topology compact sets are closed sets that are bounded away from 0.[1] We denote the Borel σ-algebra on $(\bar{\mathbb{R}}_0^d, \tau_0)$ by $\mathfrak{B}(\bar{\mathbb{R}}_0^d)$, i.e. $\mathfrak{B}(\bar{\mathbb{R}}_0^d) = \sigma(\tau_0)$.

To define convergence of a sequence in $\bar{\mathbb{R}}^d$, we first define convergence to a point $x \neq 0$ as in the previous case. For $x_1, x_2, \cdots \in \bar{\mathbb{R}}^d$ we write $x_n \to 0$ or $\lim_{n\to\infty} x_n = 0$ when $\vartheta(x_n) \stackrel{\bar{\mathbb{R}}_+}{\to} 0$. Note that if $x, x_1, x_2, \cdots \in \bar{\mathbb{R}}^d \setminus \mathbb{I}^{d-1}$ then $x_n \to x$ if and only if $x_n \stackrel{\mathbb{R}^d}{\to} x$. Let τ be the topology induced by this definition of convergence, and let $\mathfrak{B}(\bar{\mathbb{R}}^d) = \sigma(\tau)$ be the Borel σ-algebra on $\bar{\mathbb{R}}^d$ with this topology. In this space the compact sets are just the closed sets.

[1] A set A is said to be **bounded away from** 0 if 0 is not in the closure of A, i.e. $0 \notin \bar{A}$.

For notational convenience, we identify Borel measures on $\bar{\mathbb{R}}^d_0$ with Borel measures on $\bar{\mathbb{R}}^d$ that place no mass at zero. Likewise, we identify Borel measures on $\mathbb{R}^d$ with Borel measures on $\bar{\mathbb{R}}^d$ that place no mass on $\mathbb{I}^{d-1}$.

4.2.2 Vague Convergence

Although we are mainly interested in vague convergence of measures on $\mathbb{R}^d$ and its compactifications, we will need several results from the general theory. Let $\mathbb{E}$ be a set equipped with a topology $\mathscr{T}$. Throughout, we assume that $(\mathbb{E}, \mathscr{T})$ is a locally compact Hausdorff space with a countable basis. By locally compact we mean that every $x \in \mathbb{E}$ is contained in a relatively compact open set, and by countable basis we mean that there exists a countable collection of open sets $\{G_n\}$ such that every open set G can be written as a finite or countable union of elements in $\{G_n\}$. By Theorem 7.6.1 in [8] this implies that $\mathbb{E}$ is a Polish space and is thus metrizable as a complete and separable metric space. As usual, we denote the Borel σ-algebra by $\mathfrak{B}(\mathbb{E}) = \sigma(\mathscr{T})$. A Borel measure on $\mathbb{E}$ is called a **Radon measure** if it is finite on any compact subset of $\mathbb{E}$. We denote the space of all Radon measures on $\mathbb{E}$ by $M(\mathbb{E})$.

Definition 4.2. If $\mu, \mu_1, \mu_2, \cdots \in M(\mathbb{E})$, we say that $\{\mu_n\}$ **converges vaguely** to μ on $\mathbb{E}$ and write $\mu_n \stackrel{v}{\rightarrow} \mu$ on $\mathbb{E}$ if for any continuous, real-valued function f on $\mathbb{E}$ vanishing outside of some compact set

$$\lim_{n\to\infty} \int_{\mathbb{E}} f(x)\mu_n(\mathrm{d}x) = \int_{\mathbb{E}} f(x)\mu(\mathrm{d}x). \tag{4.7}$$

When working with vague convergence, the following definition is useful.

Definition 4.3. Let μ be a Borel measure on $\mathbb{E}$. A set $B \in \mathfrak{B}(\mathbb{E})$ is called a **continuity set** of μ if $\mu(\partial B) = 0$.

One of the most important results about vague convergence is the so-called Portmanteau Theorem. Its proof is given in, e.g., [62].

Proposition 4.4. *Let $\mu, \mu_1, \mu_2, \ldots$ be a sequence in $M(\mathbb{E})$. The following are equivalent:*

1. *$\mu_n \stackrel{v}{\rightarrow} \mu$ on $\mathbb{E}$;*
2. *$\mu_n(B) \to \mu(B)$ for all relatively compact continuity sets B of μ;*
3. *$\limsup_{n\to\infty} \mu_n(K) \le \mu(K)$ and $\liminf_{n\to\infty} \mu_n(G) \ge \mu(G)$ for all compact sets K and all open, relatively compact sets G;*
4. *for all relatively compact continuity sets B of μ and all measurable functions f, which are continuous and bounded on B, $\int_B f(x)\mu_n(\mathrm{d}x) \to \int_B f(x)\mu(\mathrm{d}x)$.*

Condition 2 of Proposition 4.4 is often the easiest to use, but showing convergence on all relatively compact continuity sets may still be quite difficult. It turns out that in many cases it suffices to show convergence only on certain simpler collections of sets.

Proposition 4.5. *Let $\mu, \mu_1, \mu_2, \ldots$ be a sequence in $M(\mathbb{E})$ and let $\mathscr{A} \subset \mathfrak{B}(\mathbb{E})$ be a class of relatively compact open sets satisfying:*

1) $\mathscr{A}$ is closed under finite intersections, and
2) any relatively compact open set is a countable union of elements of $\mathscr{A}$.

If

$$\lim_{n\to\infty} \mu_n(A) = \mu(A)$$

for every $A \in \mathscr{A}$, then $\mu_n \stackrel{v}{\to} \mu$.

Proof. If $A, B \in \mathscr{A}$, then by condition 1) $A \cap B \in \mathscr{A}$ and

$$\begin{aligned}\mu_n(A \cup B) &= \mu_n(A) + \mu_n(B) - \mu_n(A \cap B)\\ &\to \mu(A) + \mu(B) - \mu(A \cap B) = \mu(A \cup B).\end{aligned}$$

By induction it follows that for any $m \in \mathbb{N}$ if $A_1, A_2, \ldots, A_m \in \mathscr{A}$ then

$$\lim_{n\to\infty} \mu_n\left(\bigcup_{i=1}^m A_i\right) = \mu\left(\bigcup_{i=1}^m A_i\right).$$

Now, let $G \in \mathfrak{B}(\mathbb{E})$ be a relatively compact open set. By condition 2) there is a sequence $A_1, A_2, \cdots \in \mathscr{A}$ with $G = \bigcup_{i=1}^\infty A_i$. Thus, for any $\epsilon > 0$ there exists an $m \in \mathbb{N}$ such that

$$\mu(G) - \epsilon \le \mu\left(\bigcup_{i=1}^m A_i\right).$$

It follows that,

$$\mu(G) - \epsilon \le \mu\left(\bigcup_{i=1}^m A_i\right) = \lim_{n\to\infty} \mu_n\left(\bigcup_{i=1}^m A_i\right) \le \liminf_{n\to\infty} \mu_n\left(\bigcup_{i=1}^\infty A_i\right) = \liminf_{n\to\infty} \mu_n(G),$$

and hence

$$\mu(G) \le \liminf_{n\to\infty} \mu_n(G).$$

Now, let $K \in \mathfrak{B}(\mathbb{E})$ be a compact set. Since $\mathbb{E}$ is locally compact, there exists an open cover of K made up of relatively compact sets. Further, by condition 2) there is an open cover of K made up of sets in $\mathscr{A}$. Since K is compact this means that there is a finite cover of K made up of elements of $\mathscr{A}$. In other words, there exist an $m \in \mathbb{N}$ and $A_1, A_2, \ldots, A_m \in \mathscr{A}$ such that $K \subset \bigcup_{i=1}^m A_i =: A$. Since A

is a finite union of sets in $\mathscr{A}$ and $A \cap K^c$ is a relatively compact open set we have $\lim_{n\to\infty} \mu_n(A) = \mu(A)$ and $\liminf_{n\to\infty} \mu_n(A \cap K^c) \geq \mu(A \cap K^c)$. Thus observing that $K = A \setminus (A \cap K^c)$ gives

$$\begin{aligned}\limsup_{n\to\infty} \mu_n(K) &= \limsup_{n\to\infty} [\mu_n(A) - \mu_n(A \cap K^c)] \\ &\leq \limsup_{n\to\infty} \mu_n(A) - \liminf_{n\to\infty} \mu_n(A \cap K^c) \\ &\leq \mu(A) - \mu(A \cap K^c) = \mu(K).\end{aligned}$$

From here the result follows by Part 3 of Proposition 4.4. □

It is not difficult to see that $(\bar{\mathbb{R}}^d, \tau)$ and $(\bar{\mathbb{R}}^d_0, \tau_0)$ are locally compact Hausdorff spaces with a countable basis. On $\bar{\mathbb{R}}^d$ the class of Radon measures, $M(\bar{\mathbb{R}}^d)$, consists of all finite Borel measures. Further, if $\mu, \mu_1, \mu_2, \cdots \in M(\bar{\mathbb{R}}^d)$, then $\mu_n \stackrel{v}{\to} \mu$ on $\bar{\mathbb{R}}^d$ if and only if

$$\lim_{n\to\infty} \int_{\bar{\mathbb{R}}^d} f(x)\mu_n(\mathrm{d}x) = \int_{\bar{\mathbb{R}}^d} f(x)\mu(\mathrm{d}x) \tag{4.8}$$

for all continuous real-valued function f on $\bar{\mathbb{R}}^d$. On $\bar{\mathbb{R}}^d_0$ the class of Radon measures, $M(\bar{\mathbb{R}}^d_0)$, consists of all Borel measures that are finite on any subset that is bounded away from 0. Thus all Lévy measures and all Rosiński measures are Radon measures on $\bar{\mathbb{R}}^d_0$. Further, if $\mu, \mu_1, \mu_2, \cdots \in M(\bar{\mathbb{R}}^d_0)$, then $\mu_n \stackrel{v}{\to} \mu$ on $\bar{\mathbb{R}}^d_0$ if and only if (4.8) holds for all continuous, real-valued functions f vanishing on a neighborhood of zero.

The following result is a version of Helly's Selection Theorem; it is given in a somewhat more general form in [62].[2]

Proposition 4.6. *Let $\mu_1, \mu_2, \ldots$ be a sequence of Borel measures on $\bar{\mathbb{R}}^d$ with*

$$\sup_n \mu_n(\bar{\mathbb{R}}^d) < \infty.$$

There exist a subsequence $\{\mu_{n_k}\}$ and a finite Borel measure μ on $\bar{\mathbb{R}}^d$ such that $\mu_{n_k} \stackrel{v}{\to} \mu$ on $\bar{\mathbb{R}}^d$.

We now give a useful characterization of vague convergence on $\bar{\mathbb{R}}^d$ for the special case when none of the measures place mass on $\mathbb{I}^{d-1}$. Let C^b be the class of Borel functions mapping $\bar{\mathbb{R}}^d$ into $\mathbb{R}$, which are continuous and bounded on $\bar{\mathbb{R}}^d \setminus \mathbb{I}^{d-1}$. We make no assumption about their behavior on $\mathbb{I}^{d-1}$.

---

[2]Specifically, Propositions 3.16 and 3.17 in [62] imply that $M(\bar{\mathbb{R}}^d)$ is vaguely relatively compact and metrizable as a complete and separable metric space. From here the result follows from the fact that relative compactness and sequential relative compactness are equivalent in metrizable spaces, see, e.g., pages 4–5 in [13].

Lemma 4.7. *Let $\mu, \mu_1, \mu_2, \ldots$ be finite Borel measures on $\bar{\mathbb{R}}^d$ with $\mu(\mathbb{I}^{d-1}) = 0$ and $\mu_n(\mathbb{I}^{d-1}) = 0$ for $n = 1, 2, \ldots$. In this case $\mu_n \stackrel{v}{\to} \mu$ on $\bar{\mathbb{R}}^d$ if and only if $\int_{\bar{\mathbb{R}}^d} f(x)\mu_n(\mathrm{d}x) \to \int_{\bar{\mathbb{R}}^d} f(x)\mu(\mathrm{d}x)$ for all $f \in C^b$.*

Proof. Assume that $\mu_n \stackrel{v}{\to} \mu$ on $\bar{\mathbb{R}}^d$, let $H = \{T \in (0,\infty) : \mu(|x| = T) = 0\}$, and fix $f \in C^b$. This means that there is a K such that $|f(x)| \leq K$ for all $x \in \mathbb{R}^d$. Without loss of generality assume that $f(x) \geq 0$. From the Portmanteau Theorem (Proposition 4.4) it follows that for any $T \in H$

$$\lim_{n\to\infty} \int_{|x|\leq T} f(x)\mu_n(\mathrm{d}x) = \int_{|x|\leq T} f(x)\mu(\mathrm{d}x).$$

Thus, by the Monotone Convergence Theorem and the fact that $\mu(\mathbb{I}^{d-1}) = 0$

$$\begin{aligned}\liminf_{n\to\infty} \int_{\bar{\mathbb{R}}^d} f(x)\mu_n(\mathrm{d}x) &\geq \lim_{H\ni T\uparrow\infty} \lim_{n\to\infty} \int_{|x|\leq T} f(x)\mu_n(\mathrm{d}x) \\ &= \lim_{H\ni T\uparrow\infty} \int_{|x|\leq T} f(x)\mu(\mathrm{d}x) = \int_{\bar{\mathbb{R}}^d} f(x)\mu(\mathrm{d}x).\end{aligned}$$

Further, since $\mu(\mathbb{I}^{d-1}) = 0$, for any $\delta > 0$ there is a $T_\delta \in H$ with $\mu(|x| > T_\delta) \leq \delta/K$. Thus

$$\begin{aligned}\int_{\bar{\mathbb{R}}^d} f(x)\mu_n(\mathrm{d}x) &\leq \int_{|x|\leq T_\delta} f(x)\mu_n(\mathrm{d}x) + K\mu_n(|x| > T_\delta) \\ &\to \int_{|x|\leq T_\delta} f(x)\mu(\mathrm{d}x) + K\mu(|x| > T_\delta) \\ &\leq \int_{\bar{\mathbb{R}}^d} f(x)\mu(\mathrm{d}x) + \delta.\end{aligned}$$

Since this holds for all $\delta > 0$, $\limsup_{n\to\infty} \int_{\bar{\mathbb{R}}^d} f(x)\mu_n(\mathrm{d}x) \leq \int_{\bar{\mathbb{R}}^d} f(x)\mu(\mathrm{d}x)$, and hence

$$\lim_{n\to\infty} \int_{\bar{\mathbb{R}}^d} f(x)\mu_n(\mathrm{d}x) = \int_{\bar{\mathbb{R}}^d} f(x)\mu(\mathrm{d}x).$$

The other direction follows from the definition of vague convergence on $\bar{\mathbb{R}}^d$. □

We now recall a standard result about convergence of infinitely divisible distributions in terms of vague convergence of their Lévy measures. The following is a variant of Theorem 3.1.16 and Corollary 3.1.17 in [54]. Here and throughout convergence of matrices should be interpreted as pointwise convergence of the components.

Proposition 4.8. *Let* $\mu_n = ID(A_n, M_n, b_n)$. *If* $\mu_n \stackrel{w}{\to} \mu$, *then* $\mu = ID(A, M, b)$. *Moreover,* $\mu_n \stackrel{w}{\to} \mu$ *if and only if* $M_n \stackrel{v}{\to} M$ *on* $\bar{\mathbb{R}}^d_0$, $b_n \to b$, *and*

$$\lim_{\epsilon \downarrow 0} \lim_{n\to\infty} \left(A_n + \int_{|x|\le\epsilon} xx^T M_n(\mathrm{d}x) \right) = A. \tag{4.9}$$

The result remains true if (4.9) *is replaced by*

$$\lim_{\epsilon \downarrow 0} \liminf_{n\to\infty} \left(A_n + \int_{|x|\le\epsilon} xx^T M_n(\mathrm{d}x) \right) = \lim_{\epsilon \downarrow 0} \limsup_{n\to\infty} \left(A_n + \int_{|x|\le\epsilon} xx^T M_n(\mathrm{d}x) \right) = A. \tag{4.10}$$

Many of the situations that we are interested in have a limiting distribution with a specific structure that we can exploit. In such cases we can use the following.

Lemma 4.9. *Let* $\mu_1, \mu_2, \ldots$ *be a sequence of Radon measures on* $\bar{\mathbb{R}}^d_0$, *and let* μ *be a Radon measure on* $\bar{\mathbb{R}}^d_0$ *such that* $\mu(|x| = a) = 0$ *for every* $0 < a < \infty$ *and*

$$\mu(A) = \int_{\mathbb{S}^{d-1}} \int_{(0,\infty]} 1_A(xu)\nu(\mathrm{d}x)\sigma(\mathrm{d}u)$$

for some finite Borel measure σ *on* $\mathbb{S}^{d-1}$ *and some Borel measure* ν *on* $(0,\infty]$ *that is finite outside any neighborhood of* 0. *Then* $\mu_n \stackrel{v}{\to} \mu$ *on* $\bar{\mathbb{R}}^d_0$ *if and only if*

$$\mu_n\left(|x| > t, \xi(x) \in D\right) \to \mu\left(|x| > t, \xi(x) \in D\right) \tag{4.11}$$

for every $t \in (0,\infty)$ *and every* $D \in \mathfrak{B}(\mathbb{S}^{d-1})$ *with* $\sigma(\partial D) = 0$.

Proof. If $\mu_n \stackrel{v}{\to} \mu$ on $\bar{\mathbb{R}}^d_0$, then Part 2 of Proposition 4.4 implies that (4.11) holds. Now assume that (4.11) holds. Let $\mathscr{A}$ be the class of measurable sets such that $A \in \mathscr{A}$ if and only if A is bounded away from 0 and $\mu_n(A) \to \mu(A)$. If $A, B \in \mathscr{A}$ and $A \subset B$, then

$$\mu_n(B \setminus A) = \mu_n(B) - \mu_n(A) \to \mu(B) - \mu(A) = \mu(B \setminus A),$$

and hence $B \setminus A \in \mathscr{A}$. By assumption sets of the form

$$\left\{x \in \bar{\mathbb{R}}^d : |x| > t, \xi(x) \in D\right\} \tag{4.12}$$

for $D \in \mathfrak{B}(\mathbb{S}^{d-1})$ with $\sigma(\partial D) = 0$ and $t \in (0,\infty)$ are elements of $\mathscr{A}$. Thus so are sets of the form

$$\left\{x \in \bar{\mathbb{R}}^d : a \ge |x| > b, \xi(x) \in D\right\},$$

where $D \in \mathfrak{B}(\mathbb{S}^{d-1})$ with $\sigma(\partial D) = 0$ and $0 < b < a \le \infty$. Moreover, by continuity from above, for $t \in (0, \infty)$

$$\begin{aligned}\limsup_{n\to\infty} \mu_n\left(|x| = t, \xi(x) \in D\right) &\le \lim_{\epsilon\downarrow 0} \limsup_{n\to\infty} \mu_n\left(t \ge |x| > t-\epsilon, \xi(x) \in D\right) \\ &= \lim_{\epsilon\downarrow 0} \mu\left(t \ge |x| > t-\epsilon, \xi(x) \in D\right) = 0,\end{aligned}$$

which means that all sets of the form

$$\left\{x \in \bar{\mathbb{R}}^d : a > |x| > b, \xi(x) \in D\right\}, \tag{4.13}$$

where $D \in \mathfrak{B}(\mathbb{S}^{d-1})$ with $\sigma(\partial D) = 0$ and $0 < b < a < \infty$ are elements of $\mathscr{A}$.

Let $\mathscr{A}'$ be the class of sets that includes the empty set and all sets of the form (4.12) and (4.13) with D of the required form. We claim that $\mathscr{A}'$ satisfies the assumptions of Proposition 4.5. It is immediate that $\mathscr{A}'$ is a collection of relatively compact open sets and is closed under finite intersections. Thus Assumption 1) is satisfied. Assumption 2) follows from the facts that $\bar{\mathbb{R}}^d_0$ is separable and that for any open set G and any $x \in G$ there is a set $A \in \mathscr{A}'$ such that $x \in A \subset G$. Thus, since for any $A \in \mathscr{A}'$ we have $\mu_n(A) \to \mu(A)$, the result holds by Proposition 4.5. □

4.3 Extended Rosiński Measures

We now return to our discussion of the Lévy measures of extended p-tempered α-stable distributions. Recall that the Lévy measure of such a distribution can be given by (4.6). In this section we will put it into a form that is often easier to work with. First, let ν be a Borel measure on $\bar{\mathbb{R}}^d$ such that if $B \in \mathfrak{B}(\bar{\mathbb{R}}^d)$ then

$$\nu(B) = \begin{cases} \int_{\mathbb{R}^d} 1_B(x)\left(|x|^2 \wedge |x|^\alpha\right) R(\mathrm{d}x) & \\ \qquad + \int_{\mathbb{S}^{d-1}} 1_B(\infty u) Q_u(\{0\}) \sigma(\mathrm{d}u) & \text{if } \alpha \in (0,2) \\ \int_{\mathbb{R}^d} 1_B(x)\left(|x|^2 \wedge [1 + \log^+ |x|]\right) R(\mathrm{d}x) & \text{if } \alpha = 0 \\ \int_{\mathbb{R}^d} 1_B(x)\left(|x|^2 \wedge 1\right) R(\mathrm{d}x) & \text{if } \alpha < 0 \end{cases}. \tag{4.14}$$

Note that $\nu(\{0\}) = 0$ and that for $\alpha \le 0$ we have $\nu(\mathbb{I}^{d-1}) = 0$. Note further that by (3.14) ν is a finite measure and thus ν is a Radon measure on $\bar{\mathbb{R}}^d$. We call it the **extended Rosiński measure** of the corresponding p-tempered α-stable distribution. From ν we get R back by

$$R(\mathrm{d}x) = \begin{cases} \left(|x|^2 \wedge |x|^\alpha\right)^{-1} \nu_{|_{\mathbb{R}^d}}(\mathrm{d}x) & \text{if } \alpha \in (0,2) \\ \left(|x|^2 \wedge [1 + \log^+ |x|]\right)^{-1} \nu_{|_{\mathbb{R}^d}}(\mathrm{d}x) & \text{if } \alpha = 0 \\ \left(|x|^2 \wedge 1\right)^{-1} \nu_{|_{\mathbb{R}^d}}(\mathrm{d}x) & \text{if } \alpha < 0 \end{cases}, \tag{4.15}$$

where $\nu_{|_{\mathbb{R}^d}}$ is the restriction of ν to $\mathbb{R}^d$.

Remark 4.5. Let ν be any finite Borel measure on $\bar{\mathbb{R}}^d$ with $\nu(\{0\}) = 0$. For any $p > 0$ and $\alpha \in (0, 2)$, ν is the extended Rosiński measure of some distribution in ETS^p_α. If, in addition, $\nu(\mathbb{I}^{d-1}) = 0$, then for any $p > 0$ and $\alpha < 2$, ν is the extended Rosiński measure of some distribution in ETS^p_α.

Note that ν is uniquely determined by $R(\mathrm{d}x)$ and $Q_u(\{0\})\sigma(\mathrm{d}u)$, which, in turn, uniquely determine the Lévy measure, M, of the corresponding ETS^p_α distribution. Moreover, M uniquely determines $R(\mathrm{d}x)$ and $Q_u(\{0\})\sigma(\mathrm{d}u)$. To see this note that by (4.2) M can be uniquely decomposed into the sum of a Lévy measure of a p-tempered α-stable distribution and that of an α-stable distribution. Theorem 3.3 showed that R is uniquely determined by the Lévy measure of the p-tempered α-stable part and Remark 14.4 in [69] says that $Q_u(\{0\})\sigma(\mathrm{d}u)$ is uniquely determined by the Lévy measure of the α-stable part. This leads to the following.

Proposition 4.10. *For a fixed $\alpha < 2$ and $p > 0$, the extended Rosiński measure ν is uniquely determined by the Lévy measure of the extended p-tempered α-stable distribution.*

Definition 4.11. A distribution in ETS^p_α with Gaussian part A, extended Rosiński measure ν, and shift b is denoted by $ETS^p_\alpha(A, \nu, b)$.

We conclude this section by giving a representation for the Lévy measure of a distribution in ETS^p_α in terms of its extended Rosiński measure. Fix $\mu \in ETS^p_\alpha$ with Lévy measure M given by (4.6), and let f be any Borel function, which is integrable with respect to M. If $\alpha < 0$, then

$$\int_{\mathbb{R}^d} f(x)M(\mathrm{d}x) = \int_{\bar{\mathbb{R}}^d}\int_0^\infty f(tx)t^{-1-\alpha}e^{-t^p}\mathrm{d}t\frac{1}{1\wedge|x|^2}\nu(\mathrm{d}x), \tag{4.16}$$

if $\alpha = 0$, then

$$\int_{\mathbb{R}^d} f(x)M(\mathrm{d}x) = \int_{\bar{\mathbb{R}}^d}\int_0^\infty f(tx)t^{-1}e^{-t^p}\mathrm{d}t\frac{1}{|x|^2\wedge[1+\log^+|x|]}\nu(\mathrm{d}x), \tag{4.17}$$

and if $\alpha \in (0, 2)$, then

$$\begin{aligned}
\int_{\mathbb{R}^d} f(x)M(\mathrm{d}x) &= \int_{\mathbb{S}^{d-1}}\int_0^\infty f(tu)t^{-1-\alpha}\mathrm{d}tQ_u(\{0\})\sigma(\mathrm{d}u) \\
&\quad + \int_{\mathbb{R}^d}\int_0^\infty f(tx)t^{-1-\alpha}e^{-t^p}\mathrm{d}tR(\mathrm{d}x) \\
&= \int_{\mathbb{I}^{d-1}}\int_0^\infty f(t\xi(x))t^{-1-\alpha}e^{-(t/|x|)^p}\mathrm{d}t\nu(\mathrm{d}x) \\
&\quad + \int_{\mathbb{R}^d}\int_0^\infty f(t\xi(x))t^{-1-\alpha}e^{-(t/|x|)^p}\mathrm{d}t|x|^\alpha R(\mathrm{d}x) \\
&= \int_{\bar{\mathbb{R}}^d}\int_0^\infty f(t\xi(x))t^{-1-\alpha}\frac{e^{-(t/|x|)^p}}{1\wedge|x|^{2-\alpha}}\mathrm{d}t\nu(\mathrm{d}x).
\end{aligned} \tag{4.18}$$

4.4 Sequences of Extended Tempered Stable Distributions

We now characterize the weak limits of extended p-tempered α-stable distributions.

Theorem 4.12. *Fix $\alpha < 2$, $p > 0$, and let $\mu_n = ETS_\alpha^p(A_n, \nu_n, b_n)$. If $\mu_n \stackrel{w}{\to} \mu$, then $\mu = ETS_\alpha^p(A, \nu, b)$. Moreover, $\mu_n \stackrel{w}{\to} \mu$ if and only if $\nu_n \stackrel{v}{\to} \nu$ on $\bar{\mathbb{R}}_0^d$, $b_n \to b$, and*

$$\lim_{\epsilon \downarrow 0} \lim_{n\to\infty} \left(A_n + H_n^\epsilon\right) = A, \tag{4.19}$$

where

$$H_n^\epsilon = \int_{|x|<\sqrt{\epsilon}} \frac{xx^T}{|x|^2} \int_0^{\epsilon|x|^{-1}} t^{1-\alpha} e^{-t^p} \mathrm{d}t \nu_n(\mathrm{d}x). \tag{4.20}$$

The result remains true if (4.19) *is replaced by*

$$\lim_{\epsilon \downarrow 0} \liminf_{n\to\infty} \left(A_n + H_n^\epsilon\right) = \lim_{\epsilon \downarrow 0} \limsup_{n\to\infty} \left(A_n + H_n^\epsilon\right) = A. \tag{4.21}$$

Remark 4.6. The extended Rosiński measure does not contribute to the Gaussian part if and only if

$$\lim_{\epsilon \downarrow 0} \limsup_{n\to\infty} \mathrm{tr} H_n^\epsilon = \lim_{\epsilon \downarrow 0} \limsup_{n\to\infty} \int_{|x|<\sqrt{\epsilon}} \int_0^{\epsilon|x|^{-1}} t^{1-\alpha} e^{-t^p} \mathrm{d}t \nu_n(\mathrm{d}x) = 0. \tag{4.22}$$

Since for any $\epsilon \in (0, 1)$

$$\int_{|x|<\epsilon} \nu_n(\mathrm{d}x) \int_0^1 t^{1-\alpha} e^{-t^p} \mathrm{d}t \le \mathrm{tr} H_n^\epsilon \le \int_{|x|<\sqrt{\epsilon}} \nu_n(\mathrm{d}x) \int_0^\infty t^{1-\alpha} e^{-t^p} \mathrm{d}t,$$

(4.22) holds if and only if

$$\lim_{\epsilon \downarrow 0} \limsup_{n\to\infty} \nu_n(|x| < \epsilon) = \lim_{\epsilon \downarrow 0} \limsup_{n\to\infty} \int_{|x|<\epsilon} |x|^2 R_n(\mathrm{d}x) = 0. \tag{4.23}$$

Remark 4.7. When $\alpha \le 0$ the condition $\nu_n \stackrel{v}{\to} \nu$ on $\bar{\mathbb{R}}_0^d$ is equivalent to the condition $R_n \stackrel{v}{\to} R$ on $\bar{\mathbb{R}}_0^d$ and

$$\lim_{N\to\infty} \limsup_{n\to\infty} \int_{|x|>N} \log|x| R_n(\mathrm{d}x) = 0, \quad \text{if } \alpha = 0 \tag{4.24}$$

$$\lim_{N\to\infty} \limsup_{n\to\infty} R_n(|x| > N) = 0 \quad \text{if } \alpha < 0. \tag{4.25}$$

When $\alpha \in (0,2)$ and $\nu_n(\mathbb{I}^{d-1}) = 0$ for every n then the limit does not have an α-stable part if and only if

$$\lim_{N\to\infty} \limsup_{n\to\infty} \int_{|x|>N} |x|^\alpha R_n(\mathrm{d}x) = 0. \tag{4.26}$$

In this case, the condition $\nu_n \xrightarrow{v} \nu$ on $\bar{\mathbb{R}}^d_0$ is equivalent to the condition $R_n \xrightarrow{v} R$ on $\bar{\mathbb{R}}^d_0$ and (4.26) holds.

To facilitate the proof of Theorem 4.12, we begin with several lemmas.

Lemma 4.13. *Fix $\alpha < 2$ and $p > 0$. If $s \in \mathbb{R}$ with $|s| \le 1$, then*

$$\int_0^\infty (\cos(ts) - 1)\, t^{-1-\alpha} e^{-t^p} \mathrm{d}t \le -\frac{11}{24} s^2 \int_0^1 t^{1-\alpha} e^{-t^p} \mathrm{d}t. \tag{4.27}$$

Proof. Since $\cos(x) \le 1$ we have

$$\begin{aligned}\int_0^\infty (\cos(ts) - 1)\, t^{-1-\alpha} e^{-t^p} \mathrm{d}t &\le \int_0^1 (\cos(ts) - 1)\, t^{-1-\alpha} e^{-t^p} \mathrm{d}t \\ &\le \int_0^1 \left(\frac{s^4 t^4}{24} - \frac{s^2 t^2}{2}\right) t^{-1-\alpha} e^{-t^p} \mathrm{d}t \\ &\le -\frac{11}{24} s^2 \int_0^1 t^{1-\alpha} e^{-t^p} \mathrm{d}t,\end{aligned}$$

where the second line follows by the Taylor expansion of cosine and the remainder theorem for alternating series. □

Lemma 4.14. *Let $\{\mu_n\}$ be as in Theorem 4.12.*

1. *If $\mu_n \xrightarrow{w} \mu$ for some probability measure μ, then* $\sup \nu_n(\bar{\mathbb{R}}^d) < \infty$.
2. *If $\nu_n \xrightarrow{v} \nu$ on $\bar{\mathbb{R}}^d_0$ for some finite measure ν, then* $\sup_n \nu_n(|x| \ge \epsilon) < \infty$ *for any $\epsilon > 0$.*
3. *If* (4.21) *holds, then there exists an $\epsilon > 0$ such that* $\sup \nu_n(|x| < \epsilon) < \infty$.

Proof. We begin with the first part, assume that $\mu_n \xrightarrow{w} \mu$ and get R_n from ν_n by (4.15). Combining (2.1) and (3.12) with Lemma 4.13 implies that for $|z| \le 1$

$$\begin{aligned}|\hat{\mu}_n(z)| &\le \exp\left\{\int_{|x|\le 1} \int_0^\infty (\cos(t\langle x, z\rangle) - 1)\, t^{-1-\alpha} e^{-t^p} \mathrm{d}t R_n(\mathrm{d}x)\right\} \\ &\le \exp\left\{-\frac{11}{24} \int_0^1 t^{1-\alpha} e^{-t^p} \mathrm{d}t \int_{|x|\le 1} \langle x, z\rangle^2 R_n(\mathrm{d}x)\right\},\end{aligned}$$

where the first inequality follows by the fact that we can write μ_n as the convolution of a Gaussian, an element of TS^p_α, and (when $\alpha \in (0,2)$) an α-stable distribution. By Proposition 2.5 in [69] $|\hat{\mu}_n(z)| \to |\hat{\mu}(z)|$ uniformly on compact sets, and there exists a $b > 0$ such that $|\hat{\mu}(z)| > b$ on a neighborhood of zero. Thus, for large enough n and all z in this neighborhood with $|z| \le 1$ we have

$$b < \exp\left\{-\frac{11}{24}\int_0^1 t^{1-\alpha}e^{-t^p}\mathrm{d}t\int_{|x|\le 1}\langle x,z\rangle^2 R_n(\mathrm{d}x)\right\},$$

which implies that for all such z we have $\sup_n \int_{|x|\le 1}\langle x,z\rangle^2 R_n(\mathrm{d}x) < \infty$. Clearly, if this holds on a neighborhood of zero, it must hold for every $z \in \mathbb{R}^d$. Now observe that $|x|^2 = \sum_{i=1}^d x_i^2 = \sum_{i=1}^d \langle x, e_i\rangle^2$, where $e_i \in \mathbb{R}^d$ such that e_i has zeros in all coordinates except for a 1 in the ith position. It follows that

$$\sup_n \int_{|x|\le 1} |x|^2 R_n(\mathrm{d}x) = \sup_n \nu_n\left(|x|\le 1\right) < \infty.$$

By Proposition 4.8, μ is infinitely divisible. Let M_n be the Lévy measure of μ_n and let M be the Lévy measure of μ. Let f_1 be a nonnegative, continuous, bounded, real-valued function vanishing on a neighborhood of zero with $f_1(y) = 1$ for $|y| \ge 1$. When $\alpha \in (0,2)$ by (4.18)

$$\begin{aligned}\int_{\mathbb{R}^d} f_1(x)M_n(\mathrm{d}x) &\ge \int_{\mathbb{I}^{d-1}}\int_1^\infty t^{-1-\alpha}\mathrm{d}t\nu_n(\mathrm{d}x)\\ &\quad+\int_{\infty>|x|\ge 1}\int_1^2 t^{-1-\alpha}e^{-(t/|x|)^p}\mathrm{d}t\nu_n(\mathrm{d}x)\\ &\ge \alpha^{-1}\nu_n\left(\mathbb{I}^{d-1}\right) + e^{-2^p}\frac{2^\alpha-1}{\alpha 2^\alpha}\nu_n(\infty>|x|\ge 1)\\ &\ge e^{-2^p}\frac{2^\alpha-1}{\alpha 2^\alpha}\nu_n(|x|\ge 1).\end{aligned}$$

Similarly, when $\alpha = 0$ by (4.17)

$$\begin{aligned}\int_{\mathbb{R}^d} f_1(x)M_n(\mathrm{d}x) &= \int_{\mathbb{R}^d}\int_0^\infty f_1(xt)t^{-1}e^{-t^p}\mathrm{d}t\frac{\nu_n(\mathrm{d}x)}{|x|^2\wedge(1+\log^+|x|)}\\ &\ge e^{-e^p}\int_{|x|\ge 1}\int_{|x|^{-1}}^e t^{-1}\mathrm{d}t\frac{\nu_n(\mathrm{d}x)}{1+\log|x|} = e^{-e^p}\nu_n\left(|x|\ge 1\right),\end{aligned}$$

and when $\alpha < 0$ by (4.16)

$$\begin{aligned}\int_{\mathbb{R}^d} f_1(x)M_n(\mathrm{d}x) &= \int_{\mathbb{R}^d}\int_0^\infty f_1(xt)t^{-1-\alpha}e^{-t^p}\mathrm{d}t\frac{1}{1\wedge|x|^2}\nu_n(\mathrm{d}x)\\ &\ge \nu_n(|x|\ge 1)\int_1^\infty t^{-1-\alpha}e^{-t^p}\mathrm{d}t.\end{aligned}$$

Proposition 4.8 implies that the left side converges to $\int_{\mathbb{R}^d} f_1(x)M(\mathrm{d}x)$ in all three cases. Thus, since $\int_{\mathbb{R}^d} f_1(x)M(\mathrm{d}x) < \infty$, we have $\sup_n \nu_n\left(|x|\ge 1\right) < \infty$.

The second part follows immediately from Proposition 4.4. The third part follows from the fact that (4.21) implies that there exists an $\epsilon > 0$ such that

$$\infty > \limsup_{n\to\infty} \mathrm{tr} H_n^\epsilon = \limsup_{n\to\infty} \int_{|x|<\sqrt{\epsilon}} \int_0^{\epsilon|x|^{-1}} t^{1-\alpha} e^{-t^p} \mathrm{d}t \nu_n(\mathrm{d}x)$$

$$\geq \limsup_{n\to\infty} \int_0^{\sqrt{\epsilon}} t^{1-\alpha} e^{-t^p} \mathrm{d}t \int_{|x|<\sqrt{\epsilon}} \nu_n(\mathrm{d}x),$$

and hence $\sup_n \nu_n(|x| < \sqrt{\epsilon}) < \infty$. □

Lemma 4.15. *Let $\{\mu_n\}$ be as in Theorem 4.12 and let M_n be the Lévy measure of μ_n. If $\sup \nu_n(\bar{\mathbb{R}}^d) < \infty$, then*

$$\lim_{\epsilon\downarrow 0} \lim_{n\to\infty} \left(A_n + \int_{|x|\leq\epsilon} xx^T M_n(\mathrm{d}x) \right)$$

$$= \lim_{\epsilon\downarrow 0} \lim_{n\to\infty} \left(A_n + \int_{|x|<\sqrt{\epsilon}} \frac{xx^T}{|x|^2} \int_0^{\epsilon|x|^{-1}} t^{1-\alpha} e^{-t^p} \mathrm{d}t \nu_n(\mathrm{d}x) \right), \quad (4.28)$$

whenever at least one of the limits exists. The result remains true if we replace $\lim_{n\to\infty}$ by $\liminf_{n\to\infty}$ or $\limsup_{n\to\infty}$.

Proof. We give the proof for the case when $\alpha \in (0, 2)$ only, as the other cases are similar. We can write

$$\int_{|x|\leq\epsilon} xx^T M_n(\mathrm{d}x) = \int_{\mathbb{I}^{d-1}} \int_0^\epsilon \xi(x)[\xi(x)]^T t^{1-\alpha} \mathrm{d}t \nu_n(\mathrm{d}x)$$

$$+ \int_{\infty>|x|\geq 1} \int_0^{\epsilon|x|^{-1}} xx^T t^{1-\alpha} e^{-t^p} \mathrm{d}t |x|^{-\alpha} \nu_n(\mathrm{d}x)$$

$$+ \int_{1>|x|\geq\sqrt{\epsilon}} \int_0^{\epsilon|x|^{-1}} xx^T t^{1-\alpha} e^{-t^p} \mathrm{d}t |x|^{-2} \nu_n(\mathrm{d}x)$$

$$+ \int_{|x|<\sqrt{\epsilon}} \int_0^{\epsilon|x|^{-1}} xx^T t^{1-\alpha} e^{-t^p} \mathrm{d}t |x|^{-2} \nu_n(\mathrm{d}x)$$

$$=: I_1^{n,\epsilon} + I_2^{n,\epsilon} + I_3^{n,\epsilon} + I_4^{n,\epsilon}.$$

Set $C := \sup_n \nu_n(\bar{\mathbb{R}}^d) < \infty$ and note that

$$\lim_{\epsilon\downarrow 0} \limsup_{n\to\infty} \mathrm{tr} I_1^{n,\epsilon} \leq \lim_{\epsilon\downarrow 0} C \frac{\epsilon^{2-\alpha}}{2-\alpha} = 0,$$

$$\lim_{\epsilon\downarrow 0} \limsup_{n\to\infty} \mathrm{tr} I_2^{n,\epsilon} \leq \int_{\infty>|x|\geq 1} |x|^{2-\alpha} \int_0^{\epsilon|x|^{-1}} t^{1-\alpha} \mathrm{d}t \nu_n(\mathrm{d}x) \leq \lim_{\epsilon\downarrow 0} C \frac{\epsilon^{2-\alpha}}{2-\alpha} = 0,$$

and

$$\begin{aligned}\lim_{\epsilon\downarrow 0}\limsup_{n\to\infty}\operatorname{tr} I_3^{n,\epsilon} &\leq \lim_{\epsilon\downarrow 0}\limsup_{n\to\infty}\int_{1>|x|\geq\sqrt{\epsilon}}\int_0^{\sqrt{\epsilon}} t^{1-\alpha}\mathrm{d}t\nu_n(\mathrm{d}x)\\ &\leq \lim_{\epsilon\downarrow 0} C\frac{\epsilon^{1-\alpha/2}}{2-\alpha} = 0.\end{aligned}$$

From here the result follows immediately. □

Proof (Proof of Theorem 4.12). Throughout this proof let M_n denote the Lévy measure of μ_n.

Assume that $\mu_n \xrightarrow{w} \mu$. By Proposition 4.8 μ is infinitely divisible with some Lévy triplet (A, M, b) such that $b_n \to b$, $M_n \xrightarrow{v} M$ on $\bar{\mathbb{R}}^d_0$, and (4.9) holds. Combining this with Lemmas 4.14 and 4.15 gives (4.19) which implies (4.21). It remains to show that there is an extended Rosiński measure ν such that $\mu = ETS^p_\alpha(A, \nu, b)$ and $\nu_n \xrightarrow{v} \nu$ on $\bar{\mathbb{R}}^d_0$.

By Lemma 4.14, $\sup \nu_n(\bar{\mathbb{R}}^d) < \infty$. Thus, Proposition 4.6 implies that there is a finite Borel measure $\tilde{\nu}$ on $\bar{\mathbb{R}}^d$ and a subsequence $\{\nu_{n_j}\}$ with $\nu_{n_j} \xrightarrow{v} \tilde{\nu}$ on $\bar{\mathbb{R}}^d$. Let ν be a finite Borel measure on $\bar{\mathbb{R}}^d$ with

$$\nu(A) = \tilde{\nu}\left(A \setminus \{0\}\right), \quad A \in \mathfrak{B}(\bar{\mathbb{R}}^d).$$

Note that $\nu(\{0\}) = 0$ and that $\nu_{n_j} \xrightarrow{v} \nu$ on $\bar{\mathbb{R}}^d_0$. Let f be any continuous nonnegative function on $\bar{\mathbb{R}}^d$ such that there are $\epsilon, K > 0$ with $f(x) = 0$ whenever $|x| \leq \epsilon$ and $f(x) \leq K$ for all $x \in \bar{\mathbb{R}}^d$. For $x \in \bar{\mathbb{R}}^d$ define

$$g_\alpha(x) = \begin{cases} \int_\epsilon^\infty f(\xi(x)t)t^{-1-\alpha}\frac{e^{-(t/|x|)^p}}{1\wedge|x|^{2-\alpha}}\mathrm{d}t & \alpha \in (0,2)\\ \int_{\epsilon|x|^{-1}}^\infty f(xt)\,t^{-1}\frac{e^{-t^p}}{|x|^2\wedge\left[1+\log^+|x|\right]}\mathrm{d}t & \alpha = 0\\ \int_{\epsilon|x|^{-1}}^\infty f(xt)t^{-1-\alpha}\frac{e^{-t^p}}{|x|^2\wedge 1}\mathrm{d}t & \alpha < 0.\end{cases} \tag{4.29}$$

We will show that

$$\lim_{j\to\infty}\int_{\bar{\mathbb{R}}^d} g_\alpha(x)\nu_{n_j}(\mathrm{d}x) = \int_{\bar{\mathbb{R}}^d} g_\alpha(x)\tilde{\nu}(\mathrm{d}x). \tag{4.30}$$

Assume for the moment that this holds. Observing that $g_\alpha(0) = 0$ gives

$$\begin{aligned}\int_{\mathbb{R}^d} f(x)M(\mathrm{d}x) &= \lim_{j\to\infty}\int_{\mathbb{R}^d} f(x)M_{n_j}(\mathrm{d}x) = \lim_{j\to\infty}\int_{\bar{\mathbb{R}}^d} g_\alpha(x)\nu_{n_j}(\mathrm{d}x)\\ &= \int_{\bar{\mathbb{R}}^d} g_\alpha(x)\tilde{\nu}(\mathrm{d}x) = \int_{\bar{\mathbb{R}}^d} g_\alpha(x)\nu(\mathrm{d}x),\end{aligned}$$

which implies that M is the Lévy measure of an extended p-tempered α-stable distribution with extended Rosiński measure ν. This proves that the class ETS_α^p is closed under weak convergence. Moreover, since, by Proposition 4.10, ν is uniquely determined by M, $\nu_n \xrightarrow{v} \nu$ on $\bar{\mathbb{R}}_0^d$.

We now complete the proof of this direction by showing that (4.30) holds. By the definition of vague convergence on $\bar{\mathbb{R}}^d$ it suffices to show that for each α the function g_α is bounded and continuous. When $\alpha \in (0,2)$ the facts that $\int_\epsilon^\infty t^{-1-\alpha}\mathrm{d}t < \infty$ and that $f(\xi(x)t)\frac{e^{-(t/|x|)^p}}{1\wedge|x|^{2-\alpha}}$ is uniformly bounded for $x \in \mathbb{R}^d$ and $t \geq \epsilon$ show that g_α is bounded and, by dominated convergence, it is continuous on $\bar{\mathbb{R}}^d$. When $\alpha < 0$,

$$1_{[t>\epsilon|x|^{-1}]}f(xt)t^{-1-\alpha}\frac{e^{-t^p}}{|x|^2\wedge 1} \leq Ke^{-t^p}\left(t^{-1-\alpha}+t^{1-\alpha}\epsilon^{-2}\right),$$

which is integrable on $[0,\infty)$. Thus g_α is bounded, and by dominated convergence it is continuous on $\bar{\mathbb{R}}^d$. When $\alpha = 0$, by Lemma 4.7, it suffices to show that g_α is bounded and continuous only on $\mathbb{R}^d$, so fix $x \in \mathbb{R}^d$. If $|x| \leq 1$, then

$$1_{[t\geq\epsilon|x|^{-1}]}f(xt)t^{-1}e^{-t^p}|x|^{-2} \leq 1_{[t\geq 0]}K\epsilon^{-2}te^{-t^p},$$

which is integrable with respect to t. If $|x| > 1$ fix $\delta \in (0, |x|-1)$ and let x' be such that $|x'-x| < \delta$. Then $1 < |x'| < |x| + \delta$ and

$$1_{[t>\epsilon|x'|^{-1}]}f(x't)t^{-1}e^{-t^p}\left[1+\log|x'|\right]^{-1} \leq 1_{[t\geq\epsilon(|x|+\delta)^{-1}]}Kt^{-1}e^{-t^p},$$

which is integrable with respect to t. Thus, by dominated convergence g_0 is continuous on $\mathbb{R}^d$. To show that $g_0(x)$ is bounded, note that if $|x| \leq 1$ then, as before,

$$g_0(x) \leq K\epsilon^{-2}\int_0^\infty te^{-t^p}\mathrm{d}t < \infty,$$

and if $|x| > 1$ then

$$\begin{aligned} g_0(x) &\leq K\left[1+\log|x|\right]^{-1}\int_{\epsilon|x|^{-1}}^{\epsilon e} t^{-1}\mathrm{d}t + K\int_{\epsilon e}^\infty t^{-1}e^{-t^p}\mathrm{d}t \\ &= K + K\int_{\epsilon e}^\infty t^{-1}e^{-t^p}\mathrm{d}t. \end{aligned}$$

We now turn to the other direction. Assume that $b_n \to b$, (4.21) holds, and $\nu_n \xrightarrow{v} \nu$ on $\bar{\mathbb{R}}_0^d$. We need to show that $\mu_n \xrightarrow{w} \mu$, where $\mu = ETS_\alpha^p(A,\nu,b)$. Let M be the Lévy measure of μ. Lemma 4.14 implies that $\sup \nu_n(\bar{\mathbb{R}}) < \infty$. Thus combining (4.21) with Lemma 4.15 gives (4.10). To show that $M_n \xrightarrow{v} M$ on $\bar{\mathbb{R}}_0^d$ we will show that every subsequence has a further subsequence that does this. Let $\{n_k\}$ be any increasing sequence in $\mathbb{N}$. By Proposition 4.6 there is a subsequence $\{n_{k_j}\}$ and a finite Borel measure $\tilde{\nu}$ on $\bar{\mathbb{R}}^d$ such that $\nu_{n_{k_j}} \xrightarrow{v} \tilde{\nu}$ on $\bar{\mathbb{R}}^d$. Clearly, $\nu|_{\bar{\mathbb{R}}^d_0} = \tilde{\nu}|_{\bar{\mathbb{R}}^d_0}$, where $\nu|_{\bar{\mathbb{R}}^d_0}$ and $\tilde{\nu}|_{\bar{\mathbb{R}}^d_0}$ are the restrictions, respectively, of ν and $\tilde{\nu}$ to $\bar{\mathbb{R}}^d_0$. Let f be a continuous nonnegative function on $\bar{\mathbb{R}}^d$ satisfying the same assumptions as in the previous direction, and define g_α by (4.29). Observing that $g_\alpha(0) = 0$ gives

$$\begin{aligned}\int_{\mathbb{R}^d} f(x) M_{n_{k_j}}(\mathrm{d}x) &= \int_{\bar{\mathbb{R}}^d} g_\alpha(x)\nu_{n_{k_j}}(\mathrm{d}x)\\ &\to \int_{\bar{\mathbb{R}}^d} g_\alpha(x)\tilde{\nu}(\mathrm{d}x) = \int_{\bar{\mathbb{R}}^d} g_\alpha(x)\nu(\mathrm{d}x) = \int_{\mathbb{R}^d} f(x)M(\mathrm{d}x),\end{aligned}$$

where the convergence follows by arguments similar to the previous direction. □

4.5 Closure Properties

In this section we show that ETS^p_α is, in fact, the smallest class that contains TS^p_α and is closed under weak convergence.

Proposition 4.16. *Fix $\alpha < 2$ and $p > 0$.*

1. *If $\mu = N(0, A)$, then there is a sequence $\{\mu_n\}$ in TS^p_α with $\mu_n \xrightarrow{w} \mu$.*
2. *If $\alpha \in (0, 2)$ and $\mu = S_\alpha(\sigma, 0)$, then there is a sequence $\{\mu_n\}$ in TS^p_α with $\mu_n \xrightarrow{w} \mu$.*
3. *The class ETS^p_α is the smallest class that contains TS^p_α and is closed under weak convergence. Moreover, this class is closed under taking convolutions.*

Proof. First observe that

$$\lim_{s\to 0} \frac{e^{i\langle x,z\rangle rs} - 1 - \frac{i\langle x,z\rangle sr}{1+|xr|^2 s^2}}{s^2} = -\frac{1}{2}\langle x, z\rangle^2 r^2.$$

Let $R = N(0, cA)$, where $c = \left[\int_0^\infty r^{1-\alpha} e^{-r^p} \mathrm{d}r\right]^{-1}$. Let $X = (X_1, \ldots, X_d)^T \sim R$ and define

$$R_n(B) = n^2 \int_{\mathbb{R}^d} 1_B(xn^{-1}) R(\mathrm{d}x), \qquad B \in \mathfrak{B}(\mathbb{R}^d).$$

By (3.14) this is the Rosiński measure of some distribution in TS^p_α. If $\mu_n = TS^p_\alpha(R_n, 0)$, then the cumulant generating function of μ_n satisfies

$$
\begin{aligned}
C_{\mu_n}(z) &= \int_{\mathbb{R}^d}\int_0^\infty \left(e^{i\langle x,z\rangle r} - 1 - \frac{i\langle x,z\rangle r}{1+|x|^2r^2}\right) r^{-1-\alpha}e^{-r^p}\mathrm{d}rR_n(\mathrm{d}x)\\
&= n^2\int_{\mathbb{R}^d}\int_0^\infty \left(e^{i\langle x,z\rangle r/n} - 1 - \frac{i\langle x,z\rangle r/n}{1+|x/n|^2r^2}\right) r^{-1-\alpha}e^{-r^p}\mathrm{d}rR(\mathrm{d}x)\\
&\to -\frac{1}{2}\int_{\mathbb{R}^d}\langle x,z\rangle^2R(\mathrm{d}x)\int_0^\infty r^{1-\alpha}e^{-r^p}\mathrm{d}r\\
&= -\frac{1}{2}\sum_{i=1}^d\sum_{j=1}^d z_iz_jE[X_iX_j]c^{-1} = -\frac{1}{2}\langle z, Az\rangle = C_\mu(z),
\end{aligned}
$$

where C_μ is the cumulant generating function of μ and the third line follows by dominated convergence. This implies that the first part holds. For the second part let

$$
R(B) = \int_{\mathbb{S}^{d-1}}\int_0^\infty 1_B(ut)e^{-t}t^{-\alpha}\mathrm{d}t\sigma(\mathrm{d}u), \qquad B\in\mathfrak{B}(\mathbb{R}^d),
$$

and note that

$$
\sigma(B) = \int_{\mathbb{R}^d} 1_B\left(\frac{x}{|x|}\right)|x|^\alpha R(\mathrm{d}x), \qquad B\in\mathfrak{B}(\mathbb{S}^{d-1}).
$$

Let

$$
R_n(B) = n^{-\alpha}\int_{\mathbb{R}^d} 1_B(xn)R(\mathrm{d}x), \qquad B\in\mathfrak{B}(\mathbb{R}^d).
$$

By (3.14) this is the Rosiński measure of some distribution in TS^p_α. If $\mu_n = TS^p_\alpha(R_n, 0)$, then the cumulant generating function of μ_n satisfies

$$
\begin{aligned}
C_{\mu_n}(z) &= \int_{\mathbb{R}^d}\int_0^\infty \left(e^{i\langle x,z\rangle r} - 1 - \frac{i\langle x,z\rangle r}{1+|x|^2r^2}\right) r^{-1-\alpha}e^{-r^p}\mathrm{d}rR_n(\mathrm{d}x)\\
&= n^{-\alpha}\int_{\mathbb{R}^d}\int_0^\infty \left(e^{i\langle x,z\rangle rn} - 1 - \frac{i\langle x,z\rangle rn}{1+|xn|^2r^2}\right) r^{-1-\alpha}e^{-r^p}\mathrm{d}rR(\mathrm{d}x)\\
&= \int_{\mathbb{R}^d}\int_0^\infty \left(e^{i\langle x,z\rangle t/|x|} - 1 - \frac{i\langle x,z\rangle t/|x|}{1+t^2}\right) t^{-1-\alpha}e^{-(t|x|^{-1}n^{-1})^p}\mathrm{d}t|x|^\alpha R(\mathrm{d}x)\\
&\to \int_{\mathbb{R}^d}\int_0^\infty \left(e^{i\langle x,z\rangle t/|x|} - 1 - \frac{i\langle x,z\rangle t/|x|}{1+t^2}\right) t^{-1-\alpha}\mathrm{d}t|x|^\alpha R(\mathrm{d}x)\\
&= \int_{\mathbb{S}^{d-1}}\int_0^\infty \left(e^{i\langle u,z\rangle t} - 1 - \frac{i\langle u,z\rangle t}{1+t^2}\right) t^{-1-\alpha}\mathrm{d}t\sigma(\mathrm{d}u) = C_\mu(z),
\end{aligned}
$$

where the third line follows by the substitution $t = rn|x|$ and the fourth by dominated convergence. The third part follows from the first two, Theorem 4.12, and Remark 4.3. □

Definition 4.17. For $\alpha < 2$ and $p > 0$, a random variable is said to have an **elementary p-tempered α-stable distribution** on $\mathbb{R}^d$ if it can be written as Ux, where $x \in \mathbb{R}^d$ is a nonrandom vector and $U \sim ID(0, M, b)$ is an infinitely divisible random variable on $\mathbb{R}$ with $b \in \mathbb{R}$ and $M(\mathrm{d}t) = c1_{[t>0]}t^{-1-\alpha}e^{-t^p}\mathrm{d}t$, for some $c > 0$.

For $\lambda \in \mathbb{R}$, we have

$$\mathrm{E}e^{i\lambda U} = \exp\left\{c\int_0^\infty \left(e^{i\lambda t} - 1 - \frac{i\lambda t}{1+t^2}\right)t^{-1-\alpha}e^{-t^p}\mathrm{d}t + i\lambda b\right\}. \tag{4.31}$$

Thus for $z \in \mathbb{R}^d$

$$\begin{aligned}\mathrm{E}e^{i\langle z, Ux\rangle} &= \exp\left\{c\int_0^\infty \left(e^{i\langle z,x\rangle t} - 1 - \frac{i\langle z,x\rangle t}{1+t^2}\right)t^{-1-\alpha}e^{-t^p}\mathrm{d}t + i\langle z, xb\rangle\right\}\\ &= \exp\left\{\int_{\mathbb{R}^d}\int_0^\infty \left(e^{i\langle y,z\rangle t} - 1 - \frac{i\langle y,z\rangle t}{1+|y|^2t^2}\right)t^{-1-\alpha}e^{-t^p}\mathrm{d}tR(\mathrm{d}y) + i\langle z, xb'\rangle\right\},\end{aligned}$$

where $R(\mathrm{d}y) = c\delta_x(\mathrm{d}y)$ and $b' = b + cx(1-|x|^2)\int_0^\infty \frac{1}{(1+|x|^2t^2)(1+t^2)}t^{2-\alpha}e^{-t^p}\mathrm{d}t$. Thus, $Ux \sim TS^p_\alpha(c\delta_x, b')$. Further, a distribution μ is the distribution of a finite sum of independent elementary p-tempered α-stable random variables if and only if $\mu = TS^p_\alpha(R, b)$ with R concentrated on a finite number of points. We now show that every distribution in ETS^p_α can be approximated by such distributions.

Theorem 4.18. *Fix $\alpha < 2$ and $p > 0$. The class ETS^p_α is the smallest class of distributions closed under convolution and weak convergence and containing all elementary p-tempered α-stable distributions. In fact, $\mu \in ETS^p_\alpha$ if and only if there are probability measures $\mu_1, \mu_2, \ldots$ on $\mathbb{R}^d$ with $\mu_n \xrightarrow{w} \mu$ such that each μ_n is the distribution of the sum of a finite number of independent elementary p-tempered α-stable random variables.*

For the case $p = 1$ and $\alpha \in \{-1, 0\}$ this was shown in Theorem F of [6]. There the result followed from the properties of a certain integral representation. A similar representation for the case $\alpha < 2$ and $p > 0$ is given in [51]. However, when $\alpha \in (0, 2)$ the properties of the representation are quite different, and it appears that a proof analogous to that of [6] cannot be constructed in this case. Instead, we base our proof on Theorem 4.12.

Proof. In light of Proposition 4.16, it suffices to show that we can approximate any distribution in TS^p_α. Let $\mu = TS^p_\alpha(R, b)$ and let ν be its extended Rosiński measure. Let $\{\nu_n\}$ be any sequence of finite measures on $\bar{\mathbb{R}}^d$ with finite supports such that $\nu_n(\{0\}) = 0$, $\nu_n(\mathbb{I}^{d-1}) = 0$, and $\nu_n \xrightarrow{v} \nu$ on $\bar{\mathbb{R}}^d$ (such measures exist by, e.g., Theorem 7.7.3 in [8]). By the Portmanteau Theorem (Proposition 4.4)

$$\lim_{\epsilon \downarrow 0} \limsup_{n \to \infty} \nu_n (|x| < \epsilon) \leq \lim_{\epsilon \downarrow 0} \nu (|x| \leq \epsilon) = \nu (|x| = 0) = 0.$$

Thus, if $\mu_n = ETS_\alpha^p(0, \nu_n, b)$, then $\mu_n \xrightarrow{w} \mu$ by (4.23) and Theorem 4.12. □

Since all elementary p-tempered α-stable distributions are proper, we get the following.

Corollary 4.1. *ETS_α^p is the smallest class of distributions closed under convolution and weak convergence and containing all proper p-tempered α-stable distributions.*

Chapter 5
Multiscale Properties of Tempered Stable Lévy Processes

In this chapter we characterize the multiscale properties of p-tempered α-stable Lévy processes. Specifically, let $X = \{X_t : t \geq 0\}$ be a p-tempered α-stable Lévy process. We will show when there exist deterministic function $a_t > 0$ and $b_t \in \mathbb{R}^d$ and a random variable Y not concentrated at a point such that

$$a_t X_t - b_t \xrightarrow{d} Y \text{ as } t \to c \tag{5.1}$$

for $c \in \{0, \infty\}$. When $c = \infty$ this is called **long time behavior** and when $c = 0$ it is called **short time behavior**.

From Lemma 2.5 it follows that in both cases Y must follow some β-stable distribution. Further, by Theorem 4.12 it must have a distribution in ETS^p_α. The only β-stable distributions in ETS^p_α are those with $\beta \in [\alpha, 2]$ if $\alpha \in (0, 2)$ and those with $\beta \in (0, 2]$ if $\alpha \leq 0$. Thus, these are the only possible limiting distributions.

An important consequence of long and short time behavior is that it can be extended to convergence at the level of processes. For $h > 0$ consider the time rescaled process $X^h = \{X_{th} : t \geq 0\}$. Theorem 15.17 in [41] implies that, if (5.1) holds, then there exist processes $\tilde{X}^h \stackrel{d}{=} X^h$ such that for all $t \geq 0$

$$\sup_{s \leq t} |a_h \tilde{X}^h_s - b_h - Y_s| \xrightarrow{p} 0 \text{ as } h \to c, \tag{5.2}$$

where $\{Y_t : t \geq 0\}$ is a Lévy process with $Y_1 \stackrel{d}{=} Y$. Thus, in a sense, long time behavior corresponds to what the process looks like when we "zoom out" and short time behavior corresponds to what the process looks like when we "zoom in" on it. When the long and short time behavior of a process are different, the process is multiscaling: it behaves differently in a long time frame from how it behaves in a short time frame.

M. Grabchak, *Tempered Stable Distributions*, SpringerBriefs in Mathematics, DOI 10.1007/978-3-319-24927-8_5

5.1 Long and Short Time Behavior

In this section we characterize the long and short time behavior of tempered stable Lévy processes. The proofs are deferred until Section 5.2. First note that if $\{X_t : t \geq 0\}$ is a Lévy process with $X_1 \sim TS^p_\alpha(R, b)$, then by Proposition 3.5 for any $a_t > 0$ and $b_t \in \mathbb{R}^d$ the distribution of $a_t X_t - b_t$ is given by $TS^p_\alpha(R_t, \eta_t)$, where

$$R_t(A) = t \int_{\mathbb{R}^d} 1_A(a_t x) R(\mathrm{d}x), \qquad A \in \mathfrak{B}(\mathbb{R}^d) \tag{5.3}$$

and η_t is given by

$$ta_t b + ta_t(1 - a_t^2) \int_{\mathbb{R}^d} \int_0^\infty \frac{x|x|^2}{(1 + |x|^2 r^2 a_t^2)(1 + |x|^2 r^2)} e^{-r^p} r^{2-\alpha} \mathrm{d}r R(\mathrm{d}x) - b_t. \tag{5.4}$$

We begin with the case where the limiting distribution is β-stable with $\beta \in (0 \vee \alpha, 2)$. From Proposition 3.12 it follows that all such β-stable distributions belong to the class TS^p_α and have a Rosiński measure given by R^β_σ as in (3.19). Note that, by Theorem 4.12 and Remark 4.7, for the long (or short) time behavior of μ to be β-stable it is necessary that

$$R_t \xrightarrow{v} R^\beta_\sigma \text{ on } \bar{\mathbb{R}}^d_0 \text{ as } t \to c,$$

where $c = \infty$ (or $c = 0$). We will show that this is also sufficient and that it is equivalent to the regular variation of R at c. For $\alpha \neq 0$ a version of this result was given in [30]. Our proof, which we defer until Section 5.2, allows for the case $\alpha = 0$ and is shorter and simpler.

Theorem 5.1. *Fix $c \in \{0, \infty\}$, $\alpha < 2$, $p > 0$, $\beta \in (0 \vee \alpha, 2)$, and let $\sigma \neq 0$ be a finite Borel measure on $\mathbb{S}^{d-1}$. Let $\{X_t : t \geq 0\}$ be a p-tempered α-stable Lévy Process with $X_1 \sim TS^p_\alpha(R, b)$ and let $Y \sim S_\beta(\sigma, 0)$. There exist non-stochastic functions $a_t > 0$ and $b_t \in \mathbb{R}^d$ such that*

$$a_t X_t - b_t \xrightarrow{d} Y \text{ as } t \to c \tag{5.5}$$

if and only if $R \in RV^c_{-\beta}(\sigma)$. Moreover, in this case, $a_\bullet \in RV^c_{-1/\beta}$,

$$a_t \sim K^{1/\beta} / V^{\leftarrow}(t) \text{ as } t \to c, \tag{5.6}$$

where $K = \beta^{-1}\sigma(\mathbb{S}^{d-1})$ and $V(t) = 1/R(|x| > t)$, and $b_\bullet$ is such that, if $\eta_\bullet$ is as given by (5.4), then $\eta_t \to 0$ as $t \to c$.

We now turn to the case when $\alpha \in (0, 2)$ and the limiting stable distribution has the same index of stability as the one being tempered. In this case, instead of the Rosiński measure or the extended Rosiński measure, we prefer to work with

$$\nu^1(\mathrm{d}x) = |x|^\alpha R(\mathrm{d}x),$$

which we call the **modified Rosiński measure**. Theorem 3.3 implies that this is a finite measure if and only if R is the Rosiński measure of a proper p-tempered α-stable distribution.

Theorem 5.2. *Fix $c \in \{0, \infty\}$, $\alpha \in (0, 2)$, $p > 0$, and let $\sigma \neq 0$ be a finite Borel measure on $\mathbb{S}^{d-1}$. Let $\{X_t : t \geq 0\}$ be a p-tempered α-stable Lévy Process with $X_1 \sim TS^p_\alpha(R, b)$ and let $Y \sim S_\alpha(\sigma, 0)$. There exist non-stochastic functions $a_t > 0$ and $b_t \in \mathbb{R}^d$ such that*

$$a_t X_t - b_t \stackrel{d}{\rightarrow} Y \text{ as } t \to c \tag{5.7}$$

if and only if $\nu^1 \in RV^c_0(\sigma)$, where $\nu^1(\mathrm{d}x) = |x|^\alpha R(\mathrm{d}x)$. Moreover, in this case, $a_\bullet \in RV^c_{-1/\alpha}$ with

$$a_t \sim K^{1/\alpha}/V^{\leftarrow}(t) \text{ as } t \to c, \tag{5.8}$$

where $K = \sigma(\mathbb{S}^{d-1})$ and $V(t) = t^\alpha/\nu^1(|x| > t)$, and $b_\bullet$ is such that, if $\eta_\bullet$ is as given by (5.4), then $\eta_t \to 0$ as $t \to c$.

Combining this with facts about the domains of attraction of infinite variance stable distribution given in, e.g., [30] we get the following result, which extends Theorem 3.18.

Corollary 5.3. *Fix $\alpha \in (0, 2)$, $p > 0$, and let $\mu = TS^p_\alpha(R, b)$. If M is the Lévy measure of μ and $\nu^1(\mathrm{d}x) = |x|^\alpha R(\mathrm{d}x)$, then*

$$\mu \in RV^\infty_{-\alpha}(\sigma) \iff M \in RV^\infty_{-\alpha}(\sigma) \iff \nu^1 \in RV^\infty_0(\sigma). \tag{5.9}$$

It turns out that when $c = 0$ and X_1 has a proper p-tempered α-stable distribution the result of Theorem 5.2 always holds. In this case Theorem 3.3 implies that ν^1 is a finite measure, and hence $\nu^1 \in RV^0_0(\sigma)$ with

$$\sigma(B) = \int_{\mathbb{R}^d} 1_B\left(\frac{x}{|x|}\right)\nu^1(\mathrm{d}x) = \int_{\mathbb{R}^d} 1_B\left(\frac{x}{|x|}\right)|x|^\alpha R(\mathrm{d}x), \qquad B \in \mathfrak{B}(\mathbb{S}^{d-1}).$$

In this case

$$V(t) \sim t^\alpha/K \text{ as } t \downarrow 0$$

and by Proposition 2.6

$$a_t \sim t^{-1/\alpha} \text{ as } t \downarrow 0.$$

Thus Theorem 5.2 implies that if $Y \sim S_\alpha(\sigma, 0)$, then for properly chosen b_t

$$\lim_{t\downarrow 0}\left(t^{-1/\alpha}X_t - b_t\right) \stackrel{d}{\to} Y \text{ as } t\downarrow 0. \tag{5.10}$$

This is not surprising because by Remark 3.5 all proper p-tempered α-stable distributions with $\alpha \in (0, 2)$ belong to the class of generalized tempered stable distributions, and, for this class, results analogous to (5.10) are given in [66].

We conclude this section by turning to the case where the limiting distribution is Gaussian, i.e. where it is a β-stable distribution with $\beta = 2$.

Theorem 5.4. *Fix $c \in \{0, \infty\}$, $\alpha < 2$, $p > 0$, and let $B \neq 0$ be a symmetric nonnegative-definite matrix. Let $\{X_t : t \geq 0\}$ be a p-tempered α-stable Lévy process with $X_1 \sim TS^p_\alpha(R, b)$ and let*

$$A_t = \int_{|x|\leq t} xx^T R(\mathrm{d}x). \tag{5.11}$$

There exist non-stochastic functions $a_t > 0$ and $b_t \in \mathbb{R}^d$ such that

$$a_t X_t - b_t \stackrel{d}{\to} N(0, B) \text{ as } t \to c \tag{5.12}$$

if and only if $A_\bullet \in MRV^c_0(B/\mathrm{tr}B)$. Moreover, in this case, $a_\bullet \in RV^c_{-1/2}$ and

$$a_t \sim K^{-1/2}/V^{\leftarrow}(t) \text{ as } t \to c, \tag{5.13}$$

where $K = \int_0^\infty s^{1-\alpha}e^{-s^p}\mathrm{d}s/\mathrm{tr}B$ and $V(t) = t^2/\int_{|x|\leq t}|x|^2R(\mathrm{d}x)$, and $b_\bullet$ is such that, if $\eta_\bullet$ is as given by (5.4), then $\eta_t \to 0$ as $t \to c$.

Note that in the case $\int_{\mathbb{R}^d}|x|^2R(\mathrm{d}x) < \infty$ dominated convergence implies that $A_\bullet \in MRV^\infty_0(B/\mathrm{tr}B)$ where $B = \int_{\mathbb{R}^d} xx^T R(\mathrm{d}x)$. Combining Theorem 5.4 with facts about the domain of attraction of the multivariate Gaussian given in [29] gives the following.

Corollary 5.5. *Fix $c \in \{0, \infty\}$, let $\mu = TS^p_\alpha(R, b)$, and let M be the Lévy measure of μ. There exists a nonnegative definite matrix $B \neq 0$ with*

$$\int_{|x|\leq \bullet} xx^T R(\mathrm{d}x) \in MRV^c_0(B) \tag{5.14}$$

if and only if

$$\int_{|x|\leq \bullet} xx^T M(\mathrm{d}x) \in MRV^c_0(B). \tag{5.15}$$

Further, if $c = \infty$ and one of (5.14) or (5.15) holds, then there is a nonnegative definite matrix $B' \neq 0$ (possible different from B) such that

$$\int_{|x|\leq \bullet} xx^T \mu(dx) \in MRV_0^\infty(B').$$

5.2 Proofs

In this section we prove the results of Section 5.1. We begin with several lemmas.

Lemma 5.6. *Fix $c \in \{0, \infty\}$, let Y be a random variable whose distribution is not concentrated at a point, let $a_\bullet$ be a positive function, and let $\{X_t : t \geq 0\}$ be a Lévy process with $X_1 \sim ID(A, M, b)$ and $M \neq 0$. Assume that there exists a deterministic function $\xi_\bullet$ taking values in $\mathbb{R}^d$ such that*

$$\lim_{t\to c} a_t X_t - \xi_t \stackrel{d}{\to} Y.$$

1. *If $c = 0$, then $\lim_{t\downarrow 0} a_t = \infty$ and $a_{1/t} \sim a_{1/(t+1)}$ as $t \to \infty$.*
2. *If $c = \infty$, then $\lim_{t\to\infty} a_t = 0$ and $a_t \sim a_{t+1}$ as $t \to \infty$.*

Proof. First assume $c = 0$. Let $\ell := \liminf_{t\downarrow 0} a_t$ and assume for the sake of contradiction that $\ell < \infty$. This means that there is a sequence of positive real numbers $\{t_n\}$ converging to 0 such that $\lim_{n\to\infty} a_{t_n} = \ell$. Consider a further subsequence $\{t_{n_i}\}$ such that $\lim_{i\to\infty} \xi_{t_{n_i}}$ exists (although we allow it to be infinite). Stochastic continuity of Lévy processes implies that $X_t \stackrel{p}{\to} 0$ as $t \downarrow 0$, thus Slutsky's Theorem implies that

$$Y = \text{d-}\lim_{i\to\infty}(a_{t_{n_i}} X_{t_{n_i}} - \xi_{t_{n_i}}) \stackrel{d}{=} \ell 0 - \lim_{i\to\infty} \xi_{t_{n_i}},$$

which contradicts the assumption that the distribution of Y is not concentrated at a point. Thus $\lim_{t\downarrow 0} a_t = \infty$.

Let $C_{X_1}(\bullet)$ be the cumulant generating function of X_1. The characteristic function of $a_{1/t}X_{1/t} - \xi_{1/t}$ is $\exp\left(\frac{1}{t} C_{X_1}(a_{1/t}z) - i\langle z, \xi_{1/t}\rangle\right)$. If $\hat{\mu}_Y(z)$ is the characteristic function of Y, then

$$\begin{aligned}
\hat{\mu}_Y(z) &= \lim_{t\to\infty} \exp\left(\frac{1}{t} C_{X_1}(a_{1/t}z) - i\langle z, \xi_{1/t}\rangle\right) \\
&= \lim_{t\to\infty} \exp\left(\frac{1}{t+1} C_{X_1}(a_{1/t}z) - i\langle z, \frac{t}{t+1}\xi_{1/t}\rangle\right),
\end{aligned}$$

which implies that

$$Y \stackrel{d}{=} \operatorname*{d-lim}_{t\to\infty} \left(a_{1/t} X_{1/(t+1)} - \frac{t}{t+1} \xi_{1/t} \right)$$

$$\stackrel{d}{=} \operatorname*{d-lim}_{t\to\infty} \left(\frac{a_{1/t}}{a_{1/(t+1)}} \left(a_{1/(t+1)} X_{1/(t+1)} - \xi_{1/(t+1)} \right) + \frac{a_{1/t}}{a_{1/(t+1)}} \xi_{1/(t+1)} - \frac{t}{t+1} \xi_{1/t} \right).$$

Since $\left(a_{1/(t+1)} X_{1/(t+1)} - \xi_{1/(t+1)}\right) \stackrel{d}{\to} Y$ as $t \to \infty$, the result follows by the Convergence of Types Theorem, see, e.g., Lemma 13.10 in [69].

Now assume that $c = \infty$. Let M_t be the Lévy measure of $a_t X_t - \xi_t$ and note that $M_t(\bullet) = tM(\bullet/a_t)$. By Lemma 2.5 Y has a stable distribution. Let M' be its Lévy measure and note that $M'(|x| = s) = 0$ for all $s > 0$. From here Propositions 4.8 and 4.4 imply that for any $s > 0$

$$\lim_{t\to\infty} tM(|x| > s/a_t) = \lim_{t\to\infty} M_t(|x| > s) = M'(|x| > s) < \infty,$$

where the finiteness follows from the fact that M' is a Lévy measure. This implies that $a_t \to 0$. Now let $X' \stackrel{d}{=} X_1$ be independent of $\{X_t : t \geq 0\}$. By Slutsky's Theorem $a_t X' \stackrel{p}{\to} 0$ as $t \to \infty$ and

$$Y \stackrel{d}{=} \operatorname*{d-lim}_{t\to\infty} (a_{t+1} X_{t+1} - \xi_{t+1})$$

$$\stackrel{d}{=} \operatorname*{d-lim}_{t\to\infty} \left(a_{t+1} X_t + a_{t+1} X' - \xi_{t+1}\right)$$

$$\stackrel{d}{=} \operatorname*{d-lim}_{t\to\infty} \left(\frac{a_{t+1}}{a_t} (a_t X_t - \xi_t) + \frac{a_{t+1}}{a_t} \xi_t - \xi_{t+1} \right).$$

Combining this with the fact that $(a_t X_t - \xi_t) \stackrel{d}{\to} Y$ as $t \to \infty$ and another application of the Convergence of Types Theorem gives the result. □

Lemma 5.7. *Fix $c \in \{0, \infty\}$. Let M be a Borel measure on $\mathbb{R}^d$ satisfying* (2.2). *Fix $\alpha, \beta \geq 0$ with $\alpha + \beta \in (0, 2)$ and define $M^1(\mathrm{d}x) = |x|^\alpha M(\mathrm{d}x)$. If $M^1 \in RV^c_{-\beta}(\sigma)$ for some $\sigma \neq 0$ and*

$$M_t(D) = t \int_{\mathbb{R}^d} 1_D(a_t x) M(\mathrm{d}x), \qquad D \in \mathfrak{B}(\mathbb{R}^d),$$

where $a_t \sim k^{1/(\beta+\alpha)}/V^{\leftarrow}(t)$ for some $k > 0$ and $V(t) = t^\alpha / M^1(|x| > t)$, then

$$\lim_{s\to 0} \limsup_{t\to c} \int_{|x|\leq s} |x|^2 M_t(\mathrm{d}x) = 0.$$

Further, if for some $\eta \in [0, \beta + \alpha)$

$$\int_{|x|>1} |x|^\eta M(\mathrm{d}x) < \infty, \text{ then } \lim_{s\to\infty} \limsup_{t\to c} \int_{|x|>s} |x|^\eta M_t(\mathrm{d}x) = 0$$

and if $\alpha = 0$ *and*

$$\int_{|x|>1} \log|x| M(\mathrm{d}x) < \infty, \text{ then } \lim_{s\to\infty} \limsup_{t\to c} \int_{|x|>s} \log|x| M_t(\mathrm{d}x) = 0.$$

Note that when $c = \infty$ Proposition 2.12 implies that if $M^1 \in RV^\infty_{-\beta}(\sigma)$, then $\int_{|x|>1} |x|^\eta M(\mathrm{d}x) < \infty$ for any $\eta < \alpha + \beta$. However, a similar result does not hold when $c = 0$.

Proof. Define

$$U(u) := \int_{|x|>u} |x|^\alpha M(\mathrm{d}x) \text{ and } U^t(u) := \int_{|x|>u} |x|^\alpha M_t(\mathrm{d}x) = ta_t^\alpha U(u/a_t).$$

Note that (2.16) implies that $U \in RV^c_{-\beta}$ and (2.8) implies that $a_\bullet \in RV^c_{-1/(\beta+\alpha)}$ and hence by Proposition 2.6 $\lim_{t\to c} a_t = 1/c$. By Proposition 2.6 it follows that as $t \to c$

$$t \sim V(V^\leftarrow(t)) \sim \frac{k^{\alpha/(\alpha+\beta)}}{a_t^\alpha M^1(|x| > k^{1/(\beta+\alpha)}/a_t)} \sim \frac{k}{a_t^\alpha U(1/a_t)}.$$

Combining this with Fubini's Theorem gives

$$\begin{aligned}
\lim_{t\to c} \int_{|x|\le s} |x|^2 M_t(\mathrm{d}x) &= \lim_{t\to c} (2-\alpha) \int_{|x|\le s} \int_0^{|x|} u^{1-\alpha} \mathrm{d}u |x|^\alpha M_t(\mathrm{d}x) \\
&= \lim_{t\to c} \left[(2-\alpha) \int_0^s u^{1-\alpha} U^t(u) \mathrm{d}u - s^{2-\alpha} U^t(s) \right] \\
&= \lim_{t\to c} ta_t^\alpha \left[(2-\alpha) \int_0^s u^{1-\alpha} U(u/a_t) \mathrm{d}u - s^{2-\alpha} U(s/a_t) \right] \\
&= \lim_{t\to c} k \left[(2-\alpha) \frac{\int_0^s u^{1-\alpha} U(u/a_t) \mathrm{d}u}{U(1/a_t)} - s^{2-\alpha} \frac{U(s/a_t)}{U(1/a_t)} \right] \\
&= \lim_{t\to c} k(2-\alpha) \frac{a_t^{2-\alpha} \int_0^{s/a_t} u^{1-\alpha} U(u) \mathrm{d}u}{U(1/a_t)} - ks^{2-\alpha-\beta} \\
&= \lim_{t\to c} k(2-\alpha) \frac{\int_0^{s/a_t} u^{1-\alpha} U(u) \mathrm{d}u}{(s/a_t)^{2-\alpha} U(s/a_t)} s^{2-\alpha-\beta} - ks^{2-\alpha-\beta} \\
&= k \frac{2-\alpha}{2-\alpha-\beta} s^{2-\alpha-\beta} - ks^{2-\alpha-\beta},
\end{aligned}$$

which approaches 0 as $s \to 0$. In the above the fifth equality follows by change of variables and the seventh by Karamata's Theorem (Theorem 2.7). The proofs of the other parts are similar. We just need to note that by Fubini's Theorem for $\eta \in [0, \beta + \alpha)$ and $s > 0$ we have

$$\int_{|x|>s} |x|^\eta M_t(\mathrm{d}x) = (\eta - \alpha) \int_s^\infty u^{\eta-\alpha-1} U^t(u)\mathrm{d}u + s^{\eta-\alpha} U^t(s)$$

and for $\alpha = 0$ and $s > 1$ we have

$$\int_{|x|>s} \log |x| M_t(\mathrm{d}x) = \int_s^\infty u^{-1} U^t(u)\mathrm{d}u + U^t(s) \log(s).$$

This completes the proof. □

Proof (Proof of Theorem 5.1). Note that $a_t X_t - b_t \sim TS^p_\alpha(R_t, \eta_t)$, where R_t is given by (5.3) and η_t is given by (5.4). If (5.5) holds, then Lemma 5.6 implies that $\lim_{t\to c} a_t = 1/c$ and, by Theorem 4.12 and Remark 4.7, $\lim_{t\to c} \eta_t = 0$ and $R_t \stackrel{v}{\to} R^\beta_\sigma$ on $\bar{\mathbb{R}}^d_0$ as $t \to c$. Since, for all $b \geq 0$, $R^\beta_\sigma(|x| = b) = 0$, for any $D \in \mathfrak{B}(\mathbb{S}^{d-1})$ with $\sigma(\partial D) = 0$ the Portmanteau Theorem (Proposition 4.4) implies that

$$\begin{aligned}
\lim_{t\to c} tR\left(|x| > b/a_t, \frac{x}{|x|} \in D\right) &= \lim_{t\to c} R_t\left(|x| > b, \frac{x}{|x|} \in D\right) \\
&= R^\beta_\sigma\left(|x| > b, \frac{x}{|x|} \in D\right) \\
&= \int_D \int_b^\infty r^{-1-\beta}\mathrm{d}r\sigma(\mathrm{d}u) \\
&= \beta^{-1}\sigma(D)b^{-\beta}.
\end{aligned}$$

Thus, by Proposition 2.11, $R \in RV^\infty_{-\beta}(\sigma)$, $a_\bullet \in RV^\infty_{-1/\beta}$, and (5.6) holds.

Conversely, assume that $R \in RV^\infty_{-\beta}(\sigma)$. Let R_t be as in (5.3) and a_t as in (5.6). By Proposition 2.11, for any $b > 0$ and $D \in \mathfrak{B}(\mathbb{S}^{d-1})$ with $\sigma(\partial D) = 0$

$$\begin{aligned}
\lim_{t\to c} R_t\left(|x| > b, \frac{x}{|x|} \in D\right) &= \lim_{t\to c} tR\left(|x| > b/a_t, \frac{x}{|x|} \in D\right) \\
&= \beta^{-1}\sigma(D)b^{-\alpha} \\
&= \int_D \int_b^\infty r^{-1-\beta}\mathrm{d}r\sigma(\mathrm{d}u) \\
&= R^\beta_\sigma\left(|x| > b, \frac{x}{|x|} \in D\right).
\end{aligned}$$

Since, for all $b \geq 0$, $R_\sigma^\beta(|x| = b) = 0$ we can use Lemma 4.9 to get $R_t \xrightarrow{v} R_\sigma^\beta$ on $\bar{\mathbb{R}}_0^d$ as $t \to \infty$. From here the result follows by applying Lemma[1] 5.7 and Remarks 4.6 and 4.7. □

Proof (Proof of Theorem 5.2). By Proposition 2.11 $\nu^1 \in RV_0^c(\sigma)$ if and only if there is a function $a_\bullet$ with $\lim_{t\to c} a_t = 1/c$ such that for all $s \in (0,\infty)$ and all $D \in \mathfrak{B}(\mathbb{S}^{d-1})$ with $\sigma(\partial D) = 0$

$$\lim_{t\to c} t a_t^\alpha \nu^1 \left(|x| > s/a_t, \frac{x}{|x|} \in D \right) = \sigma(D). \tag{5.16}$$

When this holds $a_\bullet \in RV_{-1/\alpha}^\infty$ and a_t is as in (5.8). Thus, it suffices to show that (5.7) holds if and only if (5.16) holds.

Let ν be the extended Rosiński measure of X_1, let ν_Y be the extended Rosiński measure of Y, and let R_t and ν_t be, respectively, the Rosiński measure and the extended Rosiński measure of $a_t X_t - b_t$. For $D \in \mathfrak{B}(\mathbb{S}^{d-1})$ with $\sigma(\partial D) = 0$ and $s \in (0,\infty)$ define $A_D^s := \{|x| > s, \xi(x) \in D\}$ and note that $\nu_Y(\partial A_D^s) = 0$.

First assume that (5.7) holds. Lemma 5.6 implies that $\lim_{t\to c} a_t = 1/c$ and Theorem 4.12 implies that $\nu_t \xrightarrow{v} \nu_Y$ on $\bar{\mathbb{R}}_0^d$ as $t \to c$. By the Portmanteau Theorem (Proposition 4.4)

$$\lim_{t\to c} \nu_t(A_D^s) = \nu_Y(A_D^s) = \nu_Y(\infty D) = \sigma(D). \tag{5.17}$$

When $s \geq 1$

$$\begin{aligned} t a_t^\alpha \nu^1\left(A_D^{s/a_t}\right) &= t a_t^\alpha \int_{|x|>s/a_t} 1_D(\xi(x))|x|^\alpha R(\mathrm{d}x) \\ &= \int_{|x|>s} 1_D(\xi(x))|x|^\alpha R_t(\mathrm{d}x) = \nu_t(A_D^s), \end{aligned}$$

and similarly when $s \in (0,1)$

$$\begin{aligned} t a_t^\alpha \nu^1\left(A_D^{s/a_t}\right) &= t a_t^\alpha \nu^1\left(A_D^{1/a_t}\right) + t a_t^\alpha \nu^1\left(A_D^{s/a_t}\right) - t a_t^\alpha \nu^1\left(A_D^{1/a_t}\right) \\ &= \nu_t(A_D^1) + \int_{1\geq|x|>s} 1_D(\xi(x))|x|^\alpha R_t(\mathrm{d}x). \end{aligned}$$

Now observe that by (5.17) when $s \in (0,1)$ we have

$$\begin{aligned} \lim_{t\to c} \int_{1\geq|x|>s} 1_D(\xi(x))|x|^\alpha R_t(\mathrm{d}x) &\leq \lim_{t\to c} s^{-(2-\alpha)} \int_{1\geq|x|>s} 1_D(\xi(x))|x|^2 R_t(\mathrm{d}x) \\ &= \lim_{t\to c} s^{-(2-\alpha)} \left[\nu_t(A_D^s) - \nu_t(A_D^1)\right] \\ &= s^{-(2-\alpha)} \left[\sigma(D) - \sigma(D)\right] = 0. \end{aligned}$$

[1] It should be noted that the parameter α means different things in Theorem 5.1 and in Lemma 5.7.

Putting everything together implies that for any $s \in (0, \infty)$

$$\lim_{t\to c} ta_t^\alpha \nu^1\left(A_D^{s/a_t}\right) = \lim_{t\to c} \nu_t(A_D^{s\vee 1}) = \sigma(D),$$

and (5.16) holds as required.

Now assume that (5.16) holds. By Proposition 2.11 $a_\bullet$ satisfies (5.8) and $a_\bullet \in RV^c_{-1/\alpha}$. As in the previous case, for $s \geq 1$ we have

$$\nu_t(A_D^s) = ta_t^\alpha \nu^1\left(A_D^{s/a_t}\right),$$

and for $s \in (0, 1)$ we have

$$\nu_t(A_D^s) = \nu_t(A_D^1) + \nu_t(A_D^s) - \nu_t(A_D^1) = ta_t^\alpha \nu^1\left(A_D^{1/a_t}\right) + \int_{1\geq|x|>s} 1_D(\xi(x))|x|^2 R_t(\mathrm{d}x).$$

Now observe that (5.16) implies that for $s \in (0, 1)$

$$\begin{aligned}\lim_{t\to c}\int_{1\geq|x|>s} 1_D(\xi(x))|x|^2 R_t(\mathrm{d}x) &\leq \lim_{t\to c}\int_{1\geq|x|>s} 1_D(\xi(x))|x|^\alpha R_t(\mathrm{d}x)\\ &= \lim_{t\to c}\left[ta_t^\alpha \nu^1(A_D^{s/a_t}) - ta_t^\alpha \nu^1(A_D^{1/a_t})\right]\\ &= \sigma(D) - \sigma(D) = 0.\end{aligned}$$

This implies that for all $s \in (0, \infty)$

$$\lim_{t\to c} \nu_t(A_D^s) = \lim_{t\to c} ta_t^\alpha \nu^1(A_D^{(s\vee 1)/a_t}) = \sigma(D),$$

and by Lemma 4.9 it follows that $\nu_t \xrightarrow{v} \nu_Y$ on $\bar{\mathbb{R}}^d_0$ as $t \to c$. Thus we have convergence of the extended Rosiński measures. It remains to show convergence of the shifts and Gaussian parts. The convergence of the shifts is equivalent to the condition that $\eta_t \to 0$ as $t \to c$. By (4.23) the limit will have no Gaussian part so long as

$$\limsup_{t\to c} \nu_t(|x| < 1) = 0,$$

which follows immediately from Lemma 5.7. This concludes the proof. □

To prove results for convergence to the Gaussian we need a few additional Lemmas.

Lemma 5.8. *Fix* $\alpha < 2$, $p > 0$, *and let* $\{R_n\}$ *be a sequence of measures on* $\mathbb{R}^d$ *satisfying* (2.2). *If, for any* $s > 0$, $\lim_{n\to\infty} R_n(|x| > s) = 0$ *and if for some* $\kappa > 0$

$$\sup_n \int_{|x|\leq\kappa} |x|^2 R_n(\mathrm{d}x) < \infty,$$

then for any $a, b, c \in (0,\infty)$

$$\lim_{n\to\infty}\left(\int_{|x|\le a} xx^T \int_0^{c/|x|} t^{1-\alpha}e^{-t^p}\mathrm{d}tR_n(\mathrm{d}x) - \zeta\int_{|x|\le b} xx^T R_n(\mathrm{d}x)\right) = 0,$$

where $\zeta = \int_0^\infty t^{1-\alpha}e^{-t^p}\mathrm{d}t$.

Proof. Fix $\epsilon > 0$ and let $C = \sup_n \int_{|x|\le\kappa} |x|^2 R_n(\mathrm{d}x)$. By dominated convergence $\lim_{u\to\infty}\int_0^u t^{1-\alpha}e^{-t^p}\mathrm{d}t = \zeta$, which implies that there exists a $u' \in (0,\infty)$ such that if $u \ge u'$, then $\left|\zeta - \int_0^u t^{1-\alpha}e^{-t^p}\mathrm{d}t\right| < \frac{\epsilon}{C}$. Fix $a' > 0$ such that $a' < \min\{c/u', a, b, \kappa\}$. For any $1 \le i, j \le d$ we have

$$\begin{aligned}
&\left|\int_{|x|\le a} x_ix_j \int_0^{c/|x|} t^{1-\alpha}e^{-t^p}\mathrm{d}tR_n(\mathrm{d}x) - \zeta\int_{|x|\le b} x_ix_jR_n(\mathrm{d}x)\right| \\
&\quad\le \left|\int_{|x|\le a'} x_ix_j \int_0^{c/|x|} t^{1-\alpha}e^{-t^p}\mathrm{d}tR_n(\mathrm{d}x) - \zeta\int_{|x|\le a'} x_ix_jR_n(\mathrm{d}x)\right| \\
&\qquad + \left|\int_{a'<|x|\le a} x_ix_j \int_0^{c/|x|} t^{1-\alpha}e^{-t^p}\mathrm{d}tR_n(\mathrm{d}x)\right| \\
&\qquad + \left|\zeta\int_{a'<|x|\le b} x_ix_jR_n(\mathrm{d}x)\right| \\
&\quad =: A_{1,n} + A_{2,n} + A_{3,n}.
\end{aligned}$$

Further,

$$A_{2,n} \le \zeta\int_{a'<|x|\le a} |x|^2R_n(\mathrm{d}x) \le \zeta a^2R_n(|x|>a') \to 0,$$

$$A_{3,n} \le \zeta\int_{a'<|x|\le b} |x|^2R_n(\mathrm{d}x) \le \zeta b^2R_n(|x|>a') \to 0,$$

and

$$\begin{aligned}
A_{1,n} &\le \int_{|x|\le a'} |x|^2\left|\zeta - \int_0^{c/|x|} t^{1-\alpha}e^{-t^p}\right|\mathrm{d}tR_n(\mathrm{d}x) \\
&\le \int_{|x|\le a'} |x|^2\left|\zeta - \int_0^{c/a'} t^{1-\alpha}e^{-t^p}\right|\mathrm{d}tR_n(\mathrm{d}x) \\
&< \frac{\epsilon}{C}\sup_n\int_{|x|\le a'} |x|^2R_n(\mathrm{d}x) \le \epsilon,
\end{aligned}$$

which completes the proof. □

Lemma 5.9. *Fix $c \in \{0, \infty\}$. Let M be a measure on $\mathbb{R}^d$ satisfying (2.2) and let $A_u = \int_{|x|\le u} xx^T M(dx)$. If $A_\bullet \in MRV_0^c(B)$ for some $B \neq 0$ and*

$$M_t(D) = t \int_{\mathbb{R}^d} 1_D(a_t x) M(dx), \qquad D \in \mathfrak{B}(\mathbb{R}^d),$$

where $a_t \sim k^{-1/2}/V^{\leftarrow}(t)$ for some $k > 0$ and $V(t) = t^2/\int_{|x|\le t} |x|^2 M(dx)$, then the following hold.

1. *There exists a $\delta > 0$ such that if $B_c^\delta = (0, \delta)$ when $c = 0$ and $B_c^\delta = (1/\delta, \infty)$ when $c = \infty$, then*

$$\sup_{t \in B_c^\delta} \int_{|x|\le 1} |x|^2 M_t(dx) < \infty.$$

2. *If, for $\eta \in [0, 2)$,*

$$\int_{|x|>1} |x|^\eta M(dx) < \infty, \tag{5.18}$$

then $\lim_{t\to c} \int_{|x|>s} |x|^\eta M_t(dx) = 0$ for all $s > 0$. Moreover, when $c = \infty$ (5.18) holds for every $\eta \in [0, 2)$.

3. *If $\int_{|x|>1} \log|x| M(dx) < \infty$, then $\lim_{s\to\infty} \limsup_{t\to c} \int_{|x|>s} \log|x| M_t(dx) = 0$.*

Proof. Let

$$U(u) := \int_{|x|\le u} |x|^2 M(dx) = \mathrm{tr} A_u \text{ and } U^t(u) := \int_{|x|\le u} |x|^2 M_t(dx) = t a_t^2 U(u/a_t).$$

From Definition 2.8, (2.8), and Proposition 2.6 it follows that $U \in RV_0^c$, $a_\bullet \in RV_{-1/2}^c$, $\lim_{t\to c} a_t = 1/c$, and $t \sim V(1/(a_t\sqrt{k})) = [k a_t^2 U(1/(a_t\sqrt{k}))]^{-1} \sim [k a_t^2 U(1/a_t)]^{-1}$ as $t \to c$. Part 1 follows from the fact that

$$\lim_{t\to c} \int_{|x|\le 1} |x|^2 M_t(dx) = \lim_{t\to c} t a_t^2 \int_{|x|\le 1/a_t} |x|^2 M(dx) = \lim_{t\to c} \frac{U(1/a_t)}{kU(1/a_t)} = 1/k < \infty.$$

Now to show Part 2. By Fubini's Theorem it follows that for any $s > 0$

$$\int_{|x|>s} |x|^\eta M_t(dx) = (2-\eta) \int_s^\infty u^{\eta-3} U^t(u) du - s^{\eta-2} U^t(s).$$

When $c = \infty$ the right side is finite by Lemma 2 on Page 277 in [23], and hence the left side must be finite as well. Further, we have

$$
\begin{aligned}
\lim_{t\to c}\int_{|x|>s}|x|^{\eta}M_t(\mathrm{d}x) &= \lim_{t\to c} ta_t^2\left[(2-\eta)\int_s^{\infty}u^{\eta-3}U(u/a_t)\mathrm{d}u - s^{\eta-2}U(s/a_t)\right]\\
&= \lim_{t\to c} k^{-1}\left[(2-\eta)\frac{\int_s^{\infty}u^{\eta-3}U(u/a_t)\mathrm{d}u}{U(1/a_t)} - s^{\eta-2}\frac{U(s/a_t)}{U(1/a_t)}\right]\\
&= \lim_{t\to c} k^{-1}(2-\eta)\frac{\int_{s/a_t}^{\infty}u^{\eta-3}U(u)\mathrm{d}u}{U(s/a_t)(s/a_t)^{\eta-2}}s^{\eta-2} - k^{-1}s^{\eta-2}\\
&= k^{-1}\left(s^{\eta-2}-s^{\eta-2}\right) = 0,
\end{aligned}
$$

where the third equality follows by change of variables and the fourth by Karamata's Theorem (Theorem 2.7). We now turn to Part 3. First consider the case $c=\infty$. The fact that $\log|x|\le|x|$ (see, e.g., 4.1.36 in [2]) and the result of Part 2 gives

$$
0 \le \lim_{s\to\infty}\limsup_{t\to\infty}\int_{|x|>s}\log|x|M_t(\mathrm{d}x) \le \lim_{s\to\infty}\limsup_{t\to\infty}\int_{|x|>s}|x|M_t(\mathrm{d}x) = 0.
$$

Now assume that $c=0$. In this case $a_t\to\infty$ as $t\to 0$ and we have

$$
\begin{aligned}
\limsup_{t\to 0}\int_{|x|>s}\log|x|M_t(\mathrm{d}x) &= \limsup_{t\to 0} t\int_{|x|>s/a_t}\log|xa_t|M(\mathrm{d}x)\\
&= \limsup_{t\to 0}\frac{\int_{|x|>s/a_t}\log|xa_t|M(\mathrm{d}x)}{ka_t^2U(1/a_t)}\\
&= \limsup_{t\to 0}\left[\frac{\int_{|x|>1}\log|xa_t|M(\mathrm{d}x)}{ka_t^2U(1/a_t)}\right.\\
&\qquad\left.+\frac{\int_{1\ge|x|>s/a_t}\log|xa_t|M(\mathrm{d}x)}{ka_t^2U(1/a_t)}\right]\\
&=: \limsup_{t\to 0}\left[I_1(t)+I_2(s,t)\right].
\end{aligned}
$$

Define

$$
f(u) = \frac{a+b\log|u|}{kU(1/u)}u^{-2},
$$

where

$$
a = \int_{|x|>1}\log|x|M(\mathrm{d}x) \text{ and } b = \int_{|x|>1}M(\mathrm{d}x).
$$

and note that, by assumption, $a, b \in (0, \infty)$. The fact that $U \in RV_0^0$ implies that $f \in RV_{-2}^{\infty}$ and thus by Proposition 2.6

$$\lim_{t\to 0} I_1(t) = \lim_{t\to 0} f(a_t) = \lim_{t\to\infty} f(t) = 0.$$

Using the inequality $\log|x| \le |x|$ again and Fubini's Theorem gives

$$\begin{aligned}
\int_{1\ge|x|>s/a_t} \log|xa_t|M(\mathrm{d}x) &\le a_t \int_{1\ge|x|>s/a_t} |x|M(\mathrm{d}x) \\
&= a_t \int_{s/a_t}^{\infty} u^{-2} \int_{s/a_t<|x|\le(u\wedge 1)} |x|^2 M(\mathrm{d}x)\mathrm{d}u \\
&\le a_t \int_{s/a_t}^{\infty} u^{-2} \int_{|x|\le(u\wedge 1)} |x|^2 M(\mathrm{d}x)\mathrm{d}u \\
&= a_t \int_{s/a_t}^{1} u^{-2} U(u)\mathrm{d}u + a_t U(1) \\
&= a_t \int_{1}^{a_t/s} U(1/u)\mathrm{d}u + a_t U(1),
\end{aligned}$$

where the final line follows by change of variables. This implies that

$$I_2(s,t) \le \frac{\int_1^{a_t/s} U(1/u)\mathrm{d}u}{ka_t U(1/a_t)} + \frac{U(1)}{ka_t U(1/a_t)} =: I_{21}(s,t) + I_{22}(t).$$

By Karamata's Theorem (Theorem 2.7) and the fact that $U(1/\bullet) \in RV_0^{\infty}$ we have

$$\lim_{s\to\infty} \limsup_{t\to 0} I_{21}(s,t) = \lim_{s\to\infty} \limsup_{t\to 0} \frac{\int_1^{a_t/s} U(1/u)\mathrm{d}u}{k(a_t/s)U(s/a_t)} s^{-1} = \lim_{s\to\infty} \frac{1}{k} s^{-1} = 0.$$

Finally, note that the function of u given by $\frac{U(1)}{kuU(1/u)}$ is an element of RV_{-1}^{∞}, which implies that $\lim_{t\to 0} I_{22}(t) = 0$. □

Proof (Proof of Theorem 5.4). Note that $a_t X_t - b_t \sim TS_\alpha^p(R_t, \eta_t)$, where R_t is given by (5.3) and η_t is given by (5.4). Before proceeding set $\zeta = \int_0^\infty s^{1-\alpha} e^{-s^p} \mathrm{d}s$.

First assume that $A_\bullet \in MRV_0^c(B/\mathrm{tr}B)$ and that a_t is given by (5.13). This implies that $a_\bullet \in RV_{-1/2}^c$. Further, Lemma 5.9 implies that the assumptions of Lemma 5.8 hold. Using this lemma gives

$$\begin{aligned}
\lim_{\epsilon\downarrow 0} \lim_{t\to c} \int_{|x|\le\sqrt{\epsilon}} xx^T \int_0^{\epsilon/|x|} s^{1-\alpha} e^{-s^p} \mathrm{d}s R_t(\mathrm{d}x) &= \lim_{t\to c} \zeta \int_{|x|\le 1/\sqrt{K}} xx^T R_t(\mathrm{d}x) \\
&= \zeta \lim_{t\to c} ta_t^2 \int_{|x|\le 1/(\sqrt{K}a_t)} xx^T R(\mathrm{d}x)
\end{aligned}$$

$$= \zeta \lim_{t\to c} K^{-1} \frac{\int_{|x|\le 1/(\sqrt{K}a_t)} xx^T R(\mathrm{d}x)}{\int_{|x|\le 1/(\sqrt{K}a_t)} |x|^2 R(\mathrm{d}x)}$$

$$= \mathrm{tr}B \lim_{t\to c} \frac{\int_{|x|\le 1/(\sqrt{K}a_t)} xx^T R(\mathrm{d}x)}{\int_{|x|\le 1/(\sqrt{K}a_t)} |x|^2 R(\mathrm{d}x)} = B.$$

From here the result will follow by Theorem 4.12. We just need to show that the extended Rosiński measure goes to zero, which follows by Remark 4.7 and Lemma 5.9.

Now assume that (5.12) holds. Theorem 4.12 implies that for every $s > 0$

$$\lim_{t\to c} R_t(|x| > s) = 0$$

and

$$\lim_{\epsilon\downarrow 0}\lim_{t\to c} \int_{|x|\le\sqrt{\epsilon}} xx^T \int_0^{\epsilon/|x|} s^{1-\alpha} e^{-s^p} \mathrm{d}s R_t(\mathrm{d}x) = B. \tag{5.19}$$

This means that there exist an $\epsilon' > 0$ and a $\delta > 0$ such that

$$\infty > \sup_{t\in B_c^\delta} \int_{|x|\le\sqrt{\epsilon'}} |x|^2 \int_0^{\epsilon'/|x|} s^{1-\alpha} e^{-s^p} \mathrm{d}s R_t(\mathrm{d}x)$$

$$\ge \sup_{t\in B_c^\delta} \int_{|x|\le\sqrt{\epsilon'}} |x|^2 R_t(\mathrm{d}x) \int_0^{\sqrt{\epsilon'}} s^{1-\alpha} e^{-s^p} \mathrm{d}s,$$

where $B_c^\delta = (0, \delta)$ if $c = 0$ and $B_c^\delta = (1/\delta, \infty)$ if $c = \infty$. Hence

$$\sup_{t\in B_c^\delta} \int_{|x|\le\sqrt{\epsilon'}} |x|^2 R_t(\mathrm{d}x) < \infty$$

and we can use Lemma 5.8, which combined with (5.19) tells us that for any $s > 0$

$$\zeta \lim_{t\to c} ta_t^2 \int_{|x|\le s/a_t} xx^T R(\mathrm{d}x) = \zeta \lim_{t\to c} \int_{|x|\le s} xx^T R_t(\mathrm{d}x) = B.$$

Thus, for any $s > 0$,

$$\zeta \lim_{t\to c} ta_t^2 U(s/a_t) = \mathrm{tr}B,$$

where $U(t) = \int_{|x|\leq t} |x|^2 R(\mathrm{d}x)$. Lemma 5.6 implies that the sequential criterion for regular variation of monotone functions (see Proposition 2.6) holds and thus that $U \in RV_0^c$. The fact that

$$\lim_{t\to c} \frac{\int_{|x|\leq t} xx^T R(\mathrm{d}x)}{\int_{|x|\leq t} |x|^2 R(\mathrm{d}x)} = \lim_{t\to c} \frac{\zeta t a_t^2 \int_{|x|\leq 1/a_t} xx^T R(\mathrm{d}x)}{\zeta t a_t^2 \int_{|x|\leq 1/a_t} |x|^2 R(\mathrm{d}x)} = \frac{B}{\mathrm{tr} B}$$

shows that $A_\bullet \in MRV_0^c(B/\mathrm{tr}B)$ as required. □

Chapter 6
Parametric Classes

Tempered stable distributions form a semiparametric class of models. However, for the purposes of specific applications one generally works with particular parametric subclass. In this chapter we discuss several parametric families and their properties.[1] We also discuss the problem of parameter estimation.

6.1 Smoothly Truncated Lévy Flights

Smoothly truncated Lévy flights form, what is perhaps, the simplest and most heavily used class of tempered stable distributions. These models have been known under a variety of names including classical tempered stable, KoBoL, and CGMY. In the case $p = 1$ they have been discovered and rediscovered multiple times, see, e.g., [39, 58, 77], and [48]. A detailed survey for the case when $p = 1$ and $\alpha \in (0, 1)$ is given in [49]. We begin our discussion with the definition.

For $p > 0$ and $\alpha < 2$, a distribution $TS^p_\alpha(R, b)$ on $\mathbb{R}$ is called a **smoothly truncated Lévy flight** (STLF) if

$$R(\mathrm{d}x) = c_-\ell_-^{-\alpha}\delta_{-\ell_-}(\mathrm{d}x) + c_+\ell_+^{-\alpha}\delta_{\ell_+}(\mathrm{d}x),$$

where $\ell_-, \ell_+ > 0$ and $c_-, c_+ \geq 0$. We use the notation $STLF^p_\alpha(c_-, c_+\ell_-, \ell_+, b)$ to denote this distribution. From Theorem 3.3 it follows that this is a proper p-tempered α-stable distribution with Lévy measure

$$M(\mathrm{d}x) = c_-|x|^{-1-\alpha}e^{-(|x|/\ell_-)^p}1_{x<0}\mathrm{d}x + c_+x^{-1-\alpha}e^{-(x/\ell_+)^p}1_{x>0}\mathrm{d}x.$$

[1] Additional parametrizations are explored in [46, 76], and [60].

M. Grabchak, *Tempered Stable Distributions*, SpringerBriefs in Mathematics, DOI 10.1007/978-3-319-24927-8_6

Further, when $\alpha \in (0,2)$ we can write $M(\mathrm{d}x) = q\left(|x|^p, \frac{x}{|x|}\right) L(\mathrm{d}x)$, where

$$L(\mathrm{d}x) = c_-|x|^{-1-\alpha} 1_{x<0}\mathrm{d}x + c_+ x^{-1-\alpha} 1_{x>0}$$

is the Lévy measure of the α-stable distribution being tempered and

$$q(r^p, u) = \begin{cases} e^{-(r/\ell_-)^p} & \text{if } u = -1 \\ e^{-(r/\ell_+)^p} & \text{if } u = +1 \end{cases}$$

is the tempering function.

Now fix $p > 0, \alpha < 2$, and let $X \sim STLF^p_\alpha(c_-, c_+\ell_-, \ell_+, b)$. From Theorem 3.15 it follows that for every $\beta \geq 0$

$$\mathrm{E}|X|^\beta < \infty,$$

and by Theorem 3.16 the cumulants are given by

$$\kappa_1 = \mathrm{E}[X] = b + \int_0^\infty \left(\frac{c_+\ell_+^{3-\alpha}}{1+\ell_+^2 t^2} - \frac{c_-\ell_-^{3-\alpha}}{1+\ell_-^2 t^2} \right) t^{2-\alpha} e^{-t^p} \mathrm{d}t,$$

$$\kappa_2 = \mathrm{Var}(X) = p^{-1}\Gamma\left(\frac{2-\alpha}{p}\right)\left[c_-\ell_-^{2-\alpha} + c_+\ell_+^{2-\alpha}\right],$$

and for $n \geq 3$

$$\kappa_n = p^{-1}\Gamma\left(\frac{n-\alpha}{p}\right)\left[c_-(-1)^n\ell_-^{n-\alpha} + c_+\ell_+^{n-\alpha}\right].$$

Theorem 3.17 implies that if $p \in (0,1]$ and $\theta_0 = \min\{\ell_-^{-p}, \ell_+^{-p}\}$, then

$$\mathrm{E}\left[e^{\theta|X|}\right] \begin{cases} < \infty & \text{if } \theta < \theta_0 \\ = \infty & \text{if } \theta > \theta_0 \end{cases}$$

and

$$\mathrm{E}\left[e^{\theta_0|X|}\right] \begin{cases} < \infty & \text{if } \alpha \in (0,2) \\ = \infty & \text{if } \alpha \leq 0 \end{cases}.$$

Hence the distribution has exponential tails in this case. However, by (3.22) and (3.23) whenever $p > 1$ we have

$$\mathrm{E}\left[e^{\theta|X|}\right] < \infty \text{ for all } \theta \geq 0,$$

and the distribution has lighter than exponential tails in this case.

Now let $\{X_t : t \geq 0\}$ be a Lévy process with $X_1 \sim STLF_\alpha^p(c_-, c_+, \ell_-, \ell_+, b)$. Since X_1 has a proper p-tempered α-stable distribution with a finite variance, Theorems 5.4 and 5.2 imply that this process has Gaussian long time behavior and when $\alpha \in (0, 2)$ it has α-stable short time behavior.

In the remainder of this section we focus on an important subclass, which corresponds to the case when $c_- = 0$. In this case the parameter ℓ_- is irrelevant, and for simplicity we denote $c = c_+$ and $\ell = \ell_+$. In this case the Rosiński measure is

$$R(\mathrm{d}x) = c\ell^{-\alpha}\delta_\ell(\mathrm{d}x). \tag{6.1}$$

We use the notation $TW_\alpha^p(c, \ell, b)$ to denote the distribution $TS_\alpha^p(R, b)$ when R is given by (6.1),[2] and we use the notation TW_α^p to denote the class of all distributions of this form. These distributions are closely related to the class of elementary p-tempered α-stable distributions introduced in Definition 4.17. Specifically, every elementary p-tempered α-stable distribution is the distribution of a random variable of the form Ux, where $x \in \mathbb{R}^d$ and $U \sim TW_\alpha^p(c, \ell, b)$ for some $c \geq 0$, $\ell > 0$, and $b \in \mathbb{R}$. Further, Theorem 4.18 implies that every p-tempered α-stable distributions on $\mathbb{R}^d$ is the limit, in distribution, of a sequence of linear combinations of independent random variables with distributions in TW_α^p.

The case $p = 1$ is particularly important and, for simplicity, we write $TW_\alpha = TW_\alpha^1$ and $TW_\alpha(c, \ell, b) = TW_\alpha^1(c, \ell, b)$. If $\mu = TW_\alpha(c, \ell, b)$, then Corollary 3.29 implies that the characteristic function is given by $\hat{\mu}(z) = e^{C_\mu(z) + izb_1}$, where for $z \in \mathbb{R}$

$$C_\mu(z) = \begin{cases} c\ell^{-\alpha}\Gamma(-\alpha)[(1 - iz\ell)^\alpha - 1 + i\alpha z\ell] & \alpha \neq 0, 1 \\ -c[\log(1 - iz\ell) + iz\ell] & \alpha = 0 \\ c\ell^{-1}[(1 - iz\ell)\log(1 - iz\ell) + iz\ell] & \alpha = 1 \end{cases} \tag{6.2}$$

and

$$b_1 = \int_{\mathbb{R}} x\mu(\mathrm{d}x) = b + c\ell^{3-\alpha}\int_0^\infty \frac{1}{1 + \ell^2 t^2} t^{2-\alpha} e^{-t}\mathrm{d}t. \tag{6.3}$$

In the case of $TW_0(c, \ell, b)$ with $b = c\ell - c\ell^3 \int_0^\infty \frac{1}{1+\ell^2 t^2} t^2 e^{-t}\mathrm{d}t$ the characteristic function reduces to

$$e^{-c\log(1 - iz\ell)} = (1 - iz\ell)^{-c}, \qquad z \in \mathbb{R},$$

which is the characteristic function of the gamma distribution with probability density

$$\frac{1}{\Gamma(c)\ell^c} x^{c-1} e^{-x/\ell} 1_{x>0}.$$

---

[2] We use this notation in honor of Tweedie [77], who first introduced these distributions in the case $\alpha \in (0, 1)$ and $p = 1$.

In the case $TW_{.5}(c,\ell,b)$ with $b = -.5c\ell^{1/2}\Gamma(-.5) - c\ell^{5/2}\int_0^\infty \frac{1}{1+\ell^2t^2}t^{3/2}e^{-t}dt$ the characteristic function reduces to

$$\exp\left\{c\ell^{-1/2}\Gamma(-1/2)[(1-iz\ell)^{1/2}-1]\right\} = \exp\left\{-2c\sqrt{\pi}[\sqrt{1/\ell - iz} - \sqrt{1/\ell}]\right\},$$

where we use the fact that $\Gamma(-1/2) = -2\Gamma(1/2) = -2\sqrt{\pi}$. This is the characteristic function of the inverse Gaussian distribution with probability density

$$ce^{2c\sqrt{\pi/\ell}}e^{-x/\ell-\pi c^2/x}x^{-3/2}1_{x>0},$$

see Section 4.4.2 in [21].

We now turn to the question of when the distribution of a Lévy process with marginal distributions in TW_α is absolutely continuous with respect to the distribution of the α-stable Lévy process that is being tempered. First, as in the discussion just prior to Theorem 3.25 let $X = \{X_t : t \geq 0\}$ be the canonical process on the space $\Omega = D([0,\infty),\mathbb{R})$ equipped with the σ-algebra $\mathscr{F} = \sigma(X_t : t \geq 0)$ and the right-continuous natural filtration $\mathscr{F}_t = \bigcap_{s>t}\sigma(X_u : u \leq s)$.

Fix $\alpha \in (0,2)$, $c,\ell > 0$, and $a,b \in \mathbb{R}$. Consider two probability measures P_0 and P on the space $(\Omega,\mathscr{F})$. Assume that, under P, X is a Lévy process with $X_1 \sim TW_\alpha(\ell,c,b)$ and that, under P_0, X is a Lévy process with $X_1 \sim S_\alpha(a,\sigma)$ where $\sigma(\{-1\}) = 0$ and $\sigma(\{1\}) = c$. Thus, under P_0 the Lévy measure of X_1 is $L(\mathrm{d}x) = cx^{-1-\alpha}1_{x>0}\mathrm{d}x$, and under P it is $q(x,x/|x|)L(\mathrm{d}x)$, where $q(x,1) = e^{-x/\ell}$ and $q(x,-1) = 0$. Assume that

$$b - a = c\ell^{1-\alpha}\int_0^\infty \frac{1}{1+\ell^2t^2}t^{-\alpha}\left(e^{-t}-1\right)\mathrm{d}t, \tag{6.4}$$

and note that

$$\int_{\mathbb{S}^0}\int_0^1 [1-q(r,u)]^2 r^{-1-\alpha}\mathrm{d}r\sigma(\mathrm{d}u) = c\int_0^1 \left[1-e^{-r/\ell}\right]^2 r^{-1-\alpha}\mathrm{d}r < \infty.$$

From here Theorem 3.25 implies that the measures P_0 and P are mutually absolutely continuous. From (2.3) it follows that the process X has only positive jumps P_0-a.s. thus $\Delta X_s \geq 0$ for all $s > 0$ P_0-a.s. From the Lévy-Itô decomposition (see, e.g., Proposition 3.7 in [21] or Theorem 19.2 in [69]) it follows that P_0-a.s.

$$\lim_{\epsilon\downarrow 0}\left(\sum_{\{s\in(0,t]:\Delta X_s>\epsilon}\Delta X(s) - tc\int_\epsilon^1 r^{-\alpha}\mathrm{d}r\right)$$
$$= X_t - ta - tc\int_0^\infty \left(1_{r\leq 1} - \frac{1}{1+r^2}\right)r^{-\alpha}\mathrm{d}r. \tag{6.5}$$

Further, Theorem 3.25 implies that the Radon-Nikodym process is given by

$$
\begin{aligned}
U_t &= \lim_{\epsilon \downarrow 0} \left\{ \sum_{\{s\in(0,t]:|\Delta X_s|>\epsilon} \log q\left(|\Delta X_s|, \frac{\Delta X_s}{|\Delta X_s|}\right) \right. \\
&\quad \left. + t \int_{\mathbb{S}^0} \int_{\epsilon}^{\infty} [1 - q(r,u)] r^{-\alpha-1} \mathrm{d}r \sigma(\mathrm{d}u) \right\} \\
&= \lim_{\epsilon \downarrow 0} \left\{ -\ell^{-1} \sum_{\{s\in(0,t]:\Delta X_s>\epsilon} \Delta X_s + tc \int_{\epsilon}^{\infty} [1 - e^{-r/\ell}] r^{-\alpha-1} \mathrm{d}r \right\} \\
&= -\lim_{\epsilon \downarrow 0} \left\{ \ell^{-1} \left(\sum_{\{s\in(0,t]:\Delta X_s>\epsilon} \Delta X_s - tc \int_{\epsilon}^{1} r^{-\alpha} \mathrm{d}r \right) \right. \\
&\quad \left. + tc \int_{\epsilon}^{\infty} [e^{-r/\ell} - 1 + r\ell^{-1} 1_{r\le 1}] r^{-\alpha-1} \mathrm{d}r \right\} .
\end{aligned}
$$

Applying (6.5) and dominated convergence gives

$$
\begin{aligned}
U_t &= -\ell^{-1} X_t + \ell^{-1} ta + \ell^{-1} tc \int_0^{\infty} \left(1_{r\le 1} - \frac{1}{1+r^2} \right) r^{-\alpha} \mathrm{d}r \\
&\quad - tc \int_0^{\infty} [e^{-r/\ell} - 1 + \ell^{-1} r 1_{r\le 1}] r^{-\alpha-1} \mathrm{d}r \\
&= -\ell^{-1} X_t + \ell^{-1} ta - tc \int_0^{\infty} \left[e^{-r/\ell} - 1 + \frac{r/\ell}{1+r^2} \right] r^{-\alpha-1} \mathrm{d}r.
\end{aligned}
$$

Note that Theorem 25.17 in [69] implies

$$
\log \mathrm{E}_{P_0}[e^{-X_t/\ell}] = tc \int_0^{\infty} \left[e^{-r/\ell} - 1 + \frac{r/\ell}{1+r^2} \right] r^{-\alpha-1} \mathrm{d}r - ta/\ell
$$

and hence

$$
e^{U_t} = e^{-X_t/\ell} \frac{1}{\mathrm{E}_{P_0}[e^{-X_t/\ell}]}.
$$

This means that for any Borel function F for which the expectation exists we have

$$
\mathrm{E}_P[F(X_t)] = \mathrm{E}_{P_0}\left[F(X_t) e^{-X_t/\ell}\right] \frac{1}{\mathrm{E}_{P_0}[e^{-X_t/\ell}]} \tag{6.6}
$$

and

$$
\mathrm{E}_{P_0}[F(X_t)] = \mathrm{E}_P\left[F(X_t) e^{X_t/\ell}\right] \frac{1}{\mathrm{E}_P[e^{X_t/\ell}]}. \tag{6.7}
$$

From Theorem 3.28 it follows that

$$\mathrm{E}_P\left[e^{X_t/\ell}\right] = \exp\left\{tc\ell^{-\alpha}\Gamma(2-\alpha)/\alpha + tb_1/\ell\right\},$$

and hence

$$\mathrm{E}_{P_0}\left[e^{-X_t/\ell}\right] = \frac{1}{\mathrm{E}_P\left[e^{X_t/\ell}\right]} = \exp\left\{-tc\ell^{-\alpha}\Gamma(2-\alpha)/\alpha - tb_1/\ell\right\},$$

where $b_1 = \mathrm{E}_P[X_1]$ is as given by (6.3). Noting that (6.6) and (6.7) do not depend on the structure of the space on which the process is defined gives the following.

Proposition 6.1. *Fix $\alpha \in (0,2)$. Let $\{X_t : t \geq 0\}$ be a Lévy process with $X_1 \sim TW_\alpha(\ell, c, b)$ and let $\{Y_t : t \geq 0\}$ be a Lévy process with $Y_1 \sim S_\alpha(a, \sigma)$ where $\sigma(\{-1\}) = 0$ and $\sigma(\{1\}) = c$. If a and b satisfy (6.4) and $b_1 = \mathrm{E}[X_1]$, then*

$$\mathrm{E}[F(X_t)] = \mathrm{E}\left[F(Y_t)e^{-Y_t/\ell}\right]\exp\left\{tc\ell^{-\alpha}\Gamma(2-\alpha)/\alpha + tb_1/\ell\right\}$$

and

$$\mathrm{E}[F(Y_t)] = \mathrm{E}\left[F(X_t)e^{X_t/\ell}\right]\exp\left\{-tc\ell^{-\alpha}\Gamma(2-\alpha)/\alpha - tb_1/\ell\right\}$$

for any Borel function F for which the expectation exists.

This means that we can evaluate such expectations using Monte-Carlo methods without needing to simulate $TW_\alpha(\ell, c, b)$ random variables. Instead, it suffices to simulate from a stable distribution, which can be done using techniques given in [20]. Further, Proposition 6.1 immediately gives the following.

Theorem 6.2. *If $\mu = TW_\alpha(\ell, c, b)$ with $\alpha \in (0,2)$, then μ has a probability density*

$$m(x) = e^{-x/\ell}g(x)e^{c\ell^{-\alpha}\Gamma(2-\alpha)/\alpha + b_1/\ell},$$

where b_1 is given by (6.3) and $g(x)$ is the probability density of the distribution $S_\alpha(\sigma, a)$ where a is given by (6.4) and we have $\sigma(\{-1\}) = 0$ and $\sigma(\{1\}) = c$.

This result can be used to find computationally nice forms for the density $m(x)$. In particular, a computationally tractable integral representation of $g(x)$ is given in [57]. Further, several series expansions for $g(x)$ are available, see, e.g., Section 14 in [69]. In addition, this result can be used to simulate random variables with density $m(x)$ using an accept-reject algorithm, see [4] or [42].

6.2 Heavy Tails

In this section we discuss two classes of p-tempered α-stable distributions with regularly varying tails. Such tails are heavier than those of the STLF models discussed in Section 6.1, but, in general, lighter than those of α-stable distributions. Models of this type are useful for a variety of applications. In particular, it is often thought that the distributions of financial returns have regularly varying tails, see, e.g., Section 7.3 in [21].

6.2.1 *Power Tempering*

In this section we focus on the class of p-tempered α-stable distributions with Rosiński measures of the form

$$R(\mathrm{d}x) = c_-(1+|x|)^{-\nu_- -1} 1_{x<0}\mathrm{d}x + c_+(1+x)^{-\nu_+ -1} 1_{x>0}\mathrm{d}x, \tag{6.8}$$

where $c_-, c_+ \geq 0$ and $\nu_-, \nu_+ > \alpha \vee 0$. We write $PT^p_\alpha(c_-, c_+, \nu_-, \nu_+, b)$ to denote a distribution $TS^p_\alpha(R, b)$ where R is as in (6.8). From (3.2) and substitution it follows that the Lévy measure of this distribution is given by

$$M(\mathrm{d}x) = c_- q(|x|^p, -1)|x|^{-1-\alpha} 1_{x<0}\mathrm{d}x + c_+ q(x^p, 1)x^{-1-\alpha} 1_{x>0}\mathrm{d}x,$$

where

$$q(r, -1) = p^{-1}\int_0^\infty e^{-ru}\left(1+u^{-1/p}\right)^{-\nu_- -1} u^{-(1+\alpha)/p-1}\mathrm{d}u$$

and

$$q(r, 1) = p^{-1}\int_0^\infty e^{-ru}\left(1+u^{-1/p}\right)^{-\nu_+ -1} u^{-(1+\alpha)/p-1}\mathrm{d}u.$$

Moreover, it can be readily checked that when $\alpha \in (-1, 2)$ we have $\int_{\mathbb{R}} |x|^\alpha R(\mathrm{d}x) < \infty$, and hence Theorem 3.3 implies that all such distributions are proper p-tempered α-stable distributions. Now, let $X \sim PT^p_\alpha(c_-, c_+, \nu_-, \nu_+, b)$, let $\nu = \min\{\nu_-, \nu_+\}$, and fix $\beta \geq 0$. From Theorem 3.15 it follows that

$$\mathrm{E}|X|^\beta < \infty \text{ if and only if } \beta < \nu.$$

When this holds formulas for the cumulants are as follows.

Proposition 6.3. *Let $\mu = PT^p_\alpha(c_-, c_+, \nu_-, \nu_+, b)$ and let $\nu = \min\{\nu_-, \nu_+\}$. If $m \in \mathbb{N}$ such that $1 \leq m < \nu$, then the mth cumulant exists and is given by*

$$\kappa_1 = b + \int_{\mathbb{R}}\int_0^\infty x\frac{|x|^2}{1+|x|^2t^2}t^{2-\alpha}e^{-t^p}\mathrm{d}tR(\mathrm{d}x)$$

if $m = 1$ *and by*

$$\kappa_m = p^{-1}\Gamma\left(\frac{m-\alpha}{p}\right)\sum_{k=0}^{m}(-1)^k\binom{m}{k}\left(\frac{c_-}{\nu_- - k} + \frac{(-1)^m c_+}{\nu_+ - k}\right)$$

if $m \geq 2$. *In particular, when* $\nu > 2$ *the variance exists and is given by*

$$2p^{-1}\Gamma\left(\frac{2-\alpha}{p}\right)\left[\frac{c_-}{\nu_-(\nu_- - 1)(\nu_- - 2)} + \frac{c_+}{\nu_+(\nu_+ - 1)(\nu_+ - 2)}\right]$$

Proof. Theorem 3.16 gives the formula for κ_1 when it exists and tells us that for $m \in \mathbb{N}$ with $2 \leq m < \nu$ we have

$$\kappa_m = p^{-1}\Gamma\left(\frac{m-\alpha}{p}\right)\left[c_-\int_{-\infty}^{0} x^m(1+|x|)^{-\nu_- -1}\mathrm{d}x + c_+\int_0^\infty x^m(1+x)^{-\nu_+ -1}\mathrm{d}x\right].$$

Change of variables and the binomial theorem imply that

$$\begin{aligned}\int_{-\infty}^{0} x^m(1+|x|)^{-\nu_- -1}\mathrm{d}x &= (-1)^m\int_0^\infty x^m(1+x)^{-\nu_- -1}\mathrm{d}x\\ &= (-1)^m\int_0^\infty [(x+1)-1]^m(1+x)^{-\nu_- -1}\mathrm{d}x\\ &= (-1)^m\sum_{k=0}^{m}\binom{m}{k}(-1)^{m-k}\int_0^\infty (1+x)^{k-\nu_- -1}\mathrm{d}x\\ &= \sum_{k=0}^{m}\binom{m}{k}(-1)^k\frac{1}{\nu_- - k}.\end{aligned}$$

A similar argument gives

$$\int_0^\infty x^m(1+x)^{-\nu_+ -1}\mathrm{d}x = \sum_{k=0}^{m}\binom{m}{k}(-1)^{k+m}\frac{1}{\nu_+ - k},$$

which completes the proof. □

We conclude this section by characterizing the tail behavior of these distributions. We begin by noting that for $r > 0$ we have

$$R(x > r) = c_+\int_r^\infty (1+x)^{-1-\nu_+}\mathrm{d}x = \frac{c_+}{\nu_+}(1+r)^{-\nu_+} \in RV^\infty_{-\nu_+}$$

and

$$R(-x > -r) = c_-\int_r^\infty (1+x)^{-1-\nu_-}\mathrm{d}x = \frac{c_-}{\nu_-}(1+r)^{-\nu_-} \in RV^\infty_{-\nu_-}.$$

Thus $R \in RV_{-\nu}^{\infty}(\sigma)$, where $\nu = \min\{\nu_-, \nu_+\}$ and σ is a measure on $\mathbb{S}^0$ given by

$$\sigma(\{-1\}) = \frac{c_-}{c_- + c_+} 1_{\nu_- = \nu_+} + 1_{\nu_- < \nu_+}$$

and

$$\sigma(\{1\}) = \frac{c_+}{c_- + c_+} 1_{\nu_+ = \nu_-} + 1_{\nu_- > \nu_+}.$$

From here Theorem 3.18 implies that

$$PT_\alpha^p(c_-, c_+, \nu_-, \nu_+, b) \in RV_{-\nu}^{\infty}(\sigma).$$

Now let $\{X_t : t \geq 0\}$ be a Lévy process with $X_1 \sim PT_\alpha^p(c_-, c_+, \nu_-, \nu_+, b)$. Theorems 5.1 and 5.4 imply that the long time behavior is ν-stable if $\nu \in (\alpha \vee 0, 2)$ and Gaussian if $\nu \geq 2$. Further, when $\alpha \in (0, 2)$, Theorem 5.2 and the fact that X_1 has a proper p-tempered α-stable distribution imply that the short time behavior is α-stable.

6.2.2 *Gamma Tempering*

Another parametric family of tempered stable distributions with heavy tails was introduced in [76] under the name "gamma tempered stable." A **gamma tempered stable** distribution is $TS_\alpha^p(R, b)$ where R is given by

$$R(\mathrm{d}x) = c_-|x|^{-\nu_- - 1} e^{-\ell_-/|x|^p} 1_{x<0}\mathrm{d}x + c_+ x^{-\nu_+ - 1} e^{-\ell_+/x^p} 1_{x>0}\mathrm{d}x$$

with $c_-, c_+ \geq 0$, $\nu_-, \nu_+ > \alpha \vee 0$, and $\ell_-, \ell_+ > 0$. It is straightforward to check that R satisfies the required conditions to be the Rosiński measure of a p-tempered α-stable distribution, and, in fact, it is a proper p-tempered α-stable distribution for every $\alpha < 2$. We denote this distribution by

$$GT_\alpha^p(c_-, c_+, \nu_-, \nu_+, \ell_-, \ell_+, b).$$

From (3.2) and substitution it follows that the Lévy measure is given by

$$M(\mathrm{d}x) = c_- q(|x|^p, -1)|x|^{-1-\alpha} 1_{x<0}\mathrm{d}x + c_+ q(x^p, 1) x^{-1-\alpha} 1_{x>0}\mathrm{d}x,$$

where

$$q(r, -1) = (r + \ell_-)^{-(\nu_- - \alpha)/p} p^{-1} \Gamma\left(\frac{\nu_- - \alpha}{p}\right)$$

and

$$q(r,1) = (r+\ell_+)^{-(\nu_+-\alpha)/p} p^{-1} \Gamma\left(\frac{\nu_+-\alpha}{p}\right).$$

Note that, in each direction, the function q is, up to multiplication by a constant, the Laplace transform of a Gamma distribution. Hence the name "gamma tempering."

Now let $X \sim GT_\alpha^p(c_-, c_+, \nu_-, \nu_+, \ell_-, \ell_+, b)$, let $\nu = \min\{\nu_-, \nu_+\}$, and fix $\beta \geq 0$. From Theorem 3.15 it follows that

$$\mathrm{E}|X|^\beta < \infty \text{ if and only if } \beta < \nu.$$

When this holds formulas for the cumulants are as follows.

Proposition 6.4. *Let* $\mu = GT_\alpha^p(c_-, c_+, \nu_-, \nu_+, \ell_-, \ell_+, b)$ *and let* $\nu = \min\{\nu_-, \nu_+\}$. *If* $m \in \mathbb{N}$ *such that* $1 \leq m < \nu$, *then the mth cumulant exists and is given by*

$$\kappa_1 = b + \int_{\mathbb{R}} \int_0^\infty x \frac{|x|^2}{1+|x|^2 t^2} t^{2-\alpha} e^{-t^p} \mathrm{d}t R(\mathrm{d}x)$$

if $m = 1$ *and by*

$$\kappa_m = p^{-2} \Gamma\left(\frac{m-\alpha}{p}\right) \left[c_+ \ell_+^{-(\nu_+-m)/p} \Gamma\left(\frac{\nu_+-m}{p}\right) + (-1)^m c_- \ell_-^{-(\nu_--m)/p} \Gamma\left(\frac{\nu_--m}{p}\right) \right]$$

if $m \geq 2$.

Proof. Theorem 3.16 gives the formula for κ_1 when it exists and tells us that for $m \in \mathbb{N}$ with $2 \leq m < \nu$ we have

$$\kappa_m = p^{-1} \Gamma\left(\frac{m-\alpha}{p}\right) \left[c_- \int_{-\infty}^0 x^m |x|^{-\nu_--1} e^{-\ell_-/|x|^p} \mathrm{d}x + c_+ \int_0^\infty x^m x^{-\nu_+-1} e^{-\ell_+/x^p} \mathrm{d}x \right].$$

Change of variables gives

$$\begin{aligned}
\int_{-\infty}^0 x^m |x|^{-\nu_--1} e^{-\ell_-/|x|^p} \mathrm{d}x &= (-1)^m \int_0^\infty x^{m-\nu_--1} e^{-\ell_-/x^p} \mathrm{d}x \\
&= (-1)^m p^{-1} \ell_-^{(m-\nu_-)/p} \int_0^\infty x^{(\nu_--m)/p-1} e^{-x} \mathrm{d}x \\
&= (-1)^m p^{-1} \ell_-^{-(\nu_--m)/p} \Gamma\left(\frac{\nu_--m}{p}\right).
\end{aligned}$$

Similarly, we have

$$\int_0^\infty x^m x^{-\nu_+-1} e^{-\ell_+/x} \mathrm{d}x = p^{-1} \ell_+^{-(\nu_+-m)/p} \Gamma\left(\frac{\nu_+ - m}{p}\right),$$

and the result follows. □

We conclude this section by characterizing the tail behavior of these distributions. Let $f(x) = e^{-\ell/x^p}$ for some $\ell > 0$ and note that $f \in RV_0^\infty$. Thus, for any $\nu > 0$ Karamata's Theorem (Theorem 2.7) implies that

$$\int_r^\infty x^{-\nu-1} e^{-\ell/x^p} \mathrm{d}x \sim \nu^{-1} r^{-\nu} e^{-\ell/r^p} \text{ as } r \to \infty.$$

It follows that for $r > 0$ we have

$$R(x > r) = c_+ \int_r^\infty x^{-\nu_+-1} e^{-\ell_+/x^p} \mathrm{d}x \sim c_+ \nu_+^{-1} r^{-\nu_+} e^{-\ell_+/r^p} \in RV_{-\nu_+}^\infty$$

and

$$R(-x > -r) = c_- \int_r^\infty x^{-\nu_--1} e^{-\ell_-/x^p} \mathrm{d}x \sim c_- \nu_-^{-1} r^{-\nu_-} e^{-\ell_-/r^p} \in RV_{-\nu_-}^\infty.$$

Thus $R \in RV_{-\nu}^\infty(\sigma)$, where $\nu = \min\{\upsilon_-, \upsilon_+\}$ and σ is a measure on $\mathbb{S}^0$ given by

$$\sigma(\{-1\}) = \frac{c_-}{c_- + c_+} 1_{\nu_-=\nu_+} + 1_{\nu_-<\nu_+}$$

and

$$\sigma(\{1\}) = \frac{c_+}{c_- + c_+} 1_{\nu_+=\nu_-} + 1_{\nu_->\nu_+}.$$

From here Theorem 3.18 implies that

$$GT_\alpha^p(c_-, c_+, \nu_-, \nu_+, \ell_-, \ell_+, b) \in RV_{-\nu}^\infty(\sigma).$$

Now let $\{X_t : t \geq 0\}$ be a Lévy process with $X_1 \sim GT_\alpha^p(c_-, c_+, \nu_-, \nu_+, \ell_-, \ell_+, b)$. Theorems 5.1 and 5.4 imply that the long time behavior is ν-stable if $\nu \in (\alpha \vee 0, 2)$ and Gaussian if $\nu \geq 2$. Further, when $\alpha \in (0, 2)$, Theorem 5.2 and the fact that X_1 has a proper p-tempered α-stable distribution imply that the short time behavior is α-stable.

6.3 Parameter Estimation

Let $\Theta \subset \mathbb{R}^J$ for some $J \in \mathbb{N}$ and assume that $\{\mu_\theta : \theta \in \Theta\}$ is a parametric family of tempered stable distributions on $\mathbb{R}$. Depending on the situation we may assume that the parameters α and p depend on θ or that they are fixed and known. Assume that we have a sample $X_1, X_2, \ldots, X_n \stackrel{\text{iid}}{\sim} \mu_{\theta^*}$ for some (unknown) $\theta^* \in \Theta$, and that we want to use this sample to estimate θ^*. In this section we discuss the two most common approaches.

The first approach is the method of cumulant matching, which is a version of the method of moments. Assume that

$$\int_{\mathbb{R}} |x|^J \mu_\theta(\mathrm{d}x) < \infty \text{ for each } \theta \in \Theta,$$

and let $\kappa_1(\theta), \kappa_2(\theta), \ldots, \kappa_J(\theta)$ be the first J cumulants of μ_θ. These can be obtained using Theorem 3.16. Let

$$\hat{m}_j = \frac{1}{n}\sum_{\ell=1}^{n} X_\ell^j \text{ for } j = 1, 2, \ldots, K$$

be the sample moments. We convert these into sample cumulants using the following recursive formula given in [73]. Let $\hat{\kappa}_1 = \hat{m}_1$ and let

$$\hat{\kappa}_j = \hat{m}_j - \sum_{\ell=1}^{j-1} \binom{j-1}{\ell} \hat{\kappa}_{j-\ell}\hat{m}_\ell \text{ for } j = 2, \ldots, J.$$

Now solve the nonlinear system of equations

$$\kappa_j(\theta) = \hat{\kappa}_j \text{ for } j = 1, \ldots, J. \tag{6.9}$$

We denote the solution to this system $\hat{\theta}^{CM}$ and call this the cumulant matching estimator. The main issue is that there is no guarantee that (6.9) has a solution, nor that the solution (when it exists) is unique. For certain classes of STLFs the existence and uniqueness of solutions is verified in [49]. However, more general results are not yet known.

The second approach is the method of maximum likelihood. In this case, instead of making assumptions about the finiteness of certain moments, we assume that for every $\theta \in \Theta$ the distribution μ_θ has a density f_θ with respect to Lebesgue measure.[3] The maximum likelihood estimator (MLE) is given by

$$\hat{\theta}^{MLE} = \operatorname*{argmax}_{\theta \in \Theta} \prod_{\ell=1}^{n} f_\theta(X_\ell).$$

[3] One can, of course, perform maximum likelihood estimation even when the density does not exist, but we do not consider that case here.

While there are, in general, no simple formulas for the density f_θ, it can be evaluated by inverting the characteristic function, see [60] and the references therein for descriptions of numerical approaches.

We now turn to the question of consistency. From Remark 3.2 we know that all tempered stable distributions with $\alpha \in [0, 2)$ belong to the class of self-decomposable distributions, and all self-decomposable distributions that are not concentrated at a point have a density with respect to Lebesgue measure. Conditions for the existence and strong consistency of the MLE for parametric families of self-decomposable distributions are given in [31]. These can be used to verify consistency of the MLE for tempered stable distributions as well. In fact, [31] verified that the MLE is consistent for several important classes of STLFs. In particular, the following was shown.

Proposition 6.5. *Fix $\epsilon \in (0, 1)$, $K > \epsilon$, and let $\Theta = [\epsilon, 2-\epsilon] \times [\epsilon, \infty)^2 \times [\epsilon, K]^2 \times \mathbb{R}$. For $\theta \in \Theta$ we write $\theta = (\alpha, c_-, c_+, \ell_-, \ell_+, b)$ and $\mu_\theta = STLF_\alpha^1(c_-, c_+, \ell_-, \ell_+, b)$. Within the parametric family $\{\mu_\theta : \theta \in \Theta\}$ the MLE is strongly consistent.*

However, [31] showed that the MLE is not always consistent. In fact, if we take $\epsilon \in (0, .5)$ and replace Θ in the above by $\Theta = [0, 2-\epsilon] \times [\epsilon, \infty)^2 \times [\epsilon, K]^2 \times \mathbb{R}$ (note that α is no longer bounded away from 0), then the MLE will not be consistent. In fact, the likelihood function will become unbounded and the MLE will not exist.

Chapter 7
Applications

In this chapter we discuss two applications of tempered stable distributions. The first is to option pricing and the second is to mobility models. This latter application is important in a number of fields including ecology, anthropology, and computer science. We also discuss the mechanism by which tempered stable distributions appear in applications.

7.1 Option Pricing

It has been observed that standard models do not do a good job of modeling the fluctuations of financial returns. In particular, the tails of Gaussian distributions are too light, while the tails of infinite variance stable distributions are too heavy. On the other hand, tempered stable distributions, which have a tail behavior somewhere between these two, seem to do a good job, see the empirical studies in, e.g., [17] and [60]. Moreover, returns are known to exhibit multiscaling behavior where their distributions are often well approximated by infinite variance stable distributions in a small time frame and by Gaussian distributions in a large time frame, see, e.g., [35]. As we saw in Chapter 5, this is very much in keeping with the behavior of tempered stable Lévy processes. In this section we discuss option pricing when returns follow such models. We note that the related problem of risk estimation with tempered stable distributions is discussed in [28, 47], and the references therein.

Consider an economy made up of a non-dividend paying stock and a money market account with fixed interest rate $r \geq 0$. Let $\{S_t : t \geq 0\}$ be the price process of the stock. This means that at time t the price of the stock is S_t. Consider a European style option on this stock that matures at time $T > 0$ and has a payoff function H. This means that at time T the option will pay $H(S_T)$ to the holder and it pays nothing at any other time. We are interested in finding an arbitrage-free price for this option. For a general reference on option pricing we refer the reader to [21].

M. Grabchak, *Tempered Stable Distributions*, SpringerBriefs in Mathematics, DOI 10.1007/978-3-319-24927-8_7

Let $\Omega = D([0, \infty), \mathbb{R})$, let $X = \{X_t : t \geq 0\}$ be the canonical process on this space, let $\mathscr{F} = \sigma(X_s : s \geq 0)$, and let $(\mathscr{F}_t)_{t\in[0,\infty)}$ be the right-continuous natural filtration induced by the process X. See the discussion just before Theorem 3.25 for details. We assume that the stock price process is of the form

$$S_t = S_0 e^{X_t},$$

where $S_0 > 0$ is a constant. Further, we assume that the dynamics of the process X are governed by the probability measure P on the space $(\Omega, \mathscr{F})$. We call P the **physical** or **market measure**.

We assume that, under the physical measure P, the process X is a Lévy process with $X_1 \sim TS^p_\alpha(R, b)$. This implies that the log-returns have tempered stable distributions. An arbitrage-free price for the option exists if there is a probability measure Q on $(\Omega, \mathscr{F})$ that is equivalent[1] to P and such that under Q the discounted stock price process $(e^{-rt}S_t)_{t\in[0,T]}$ is a martingale, i.e. that for any $0 \leq t \leq u \leq T$ we have

$$e^{-rt}S_t = \mathrm{E}_Q\left[e^{-ru}S_u|\mathscr{F}_t\right],$$

where E_Q is the expectation taken with respect to Q. When this holds, we say that Q is a **risk-neutral** probability measure. Such measures need not exist and when they do exist they need not be unique. If such a measure exists, then an arbitrage-free price for the option exists and is given by

$$e^{-rT}\mathrm{E}_Q\left[H(S_T)\right].$$

Since there may be many risk-neutral probability measures there may be many arbitrage-free prices.

Proposition 9.9 in [21] implies that a risk-neutral probability measure Q always exists so long as, under the physical measure P, the paths of the shifted Lévy process $\{X_t - rt : t \geq 0\}$ are not strictly increasing or strictly decreasing with probability 1. From Proposition 3.24 it follows that sufficient conditions for this are

$$\alpha \geq 1 \text{ and } R \neq 0$$

and

$$\alpha < 1,\ R((0,\infty)) > 0,\ \text{and } R((-\infty,0))) > 0.$$

For simplicity, in the remainder of this section we assume that $p = 1$, $\alpha \in (0, 1)$, and that R satisfies

$$R((0,\infty)) > 0,\ R((-\infty,0)) > 0,\ \text{and } \int_{|x|\leq 1} |x|^\alpha R(\mathrm{d}x) < \infty.$$

We begin by checking when the physical measure is already risk-neutral.

[1] This means that for any $A \in \mathscr{F}$ we have $P(A) = 0$ if and only if $Q(A) = 0$.

Theorem 7.1. *The physical probability measure P is a risk-neutral measure if and only if $R(x>1)=0$ and*

$$\Gamma(-\alpha)\int_{\mathbb{R}}[(1-x)^{\alpha}-1]R(\mathrm{d}x)+b_0=r, \tag{7.1}$$

where b_0 is given by (3.35).

Proof. Proposition 3.18 in [21] implies that P is a risk-neutral measure if and only if $e^{-r}\mathrm{E}[e^{X_1}]=1$. Lemma 3.27 implies that $\mathrm{E}[e^{X_1}]<\infty$ if and only if $R(x>1)=0$. Further, when this holds, Theorem 3.28 implies that $e^{-r}\mathrm{E}[e^{X_1}]=1$ if and only if (7.1) holds. □

A simple way to get an equivalent measure is by using an Esscher transform. Let $\eta_+=\inf\{\eta\geq 0: R(x>\eta)=0\}$ and $\eta_-=\inf\{\eta\geq 0: R(x<-\eta)=0\}$. Here we take $\inf\emptyset=\infty$ as usual. Set $\Theta=\left[-\eta_-^{-1},\eta_+^{-1}\right]$ and note that, by Lemma 3.27,

$$\mathrm{E}[e^{\theta X_1}]<\infty$$

if and only if $\theta\in\Theta$. For any $\theta\in\Theta$ define a probability measure Q^θ on $(\Omega,\mathscr{F})$ by the Radon-Nikodym derivative process[2]

$$\frac{\mathrm{d}Q^\theta_{|\mathscr{F}_t}}{\mathrm{d}P_{|\mathscr{F}_t}}=\frac{e^{\theta X_t}}{\mathrm{E}[e^{\theta X_t}]}=e^{\theta X_t-t\Psi^*_\alpha(\theta)},$$

where

$$\Psi^*_\alpha(\theta)=\Gamma(-\alpha)\int_{\mathbb{R}}[(1-\theta x)^{\alpha}-1]R(\mathrm{d}x)+\theta b_0$$

and b_0 is given by (3.35). Theorem 3.28 implies that the characteristic function of X_t under Q^θ is given by

$$\mathrm{E}_{Q^\theta}\left[e^{izX_t}\right]=\mathrm{E}_P\left[e^{izX_t}e^{\theta X_t-t\Psi^*_\alpha(\theta)}\right]=e^{t\left[\Psi^*_\alpha(iz+\theta)-\Psi^*_\alpha(\theta)\right]}.$$

We have

$$\begin{aligned}&\Psi^*_\alpha(iz+\theta)-\Psi^*_\alpha(\theta)\\&\quad=\Gamma(-\alpha)\int_{\mathbb{R}}[(1-(iz+\theta)x)^{\alpha}-(1-\theta x)^{\alpha}]R(\mathrm{d}x)+(iz+\theta)b_0-\theta b_0\end{aligned}$$

[2] Although the Radon-Nikodym derivative process only defines measures on $(\Omega,\mathscr{F}_t)$ for $t\geq 0$, it, in fact, uniquely determines a probability measure on $(\Omega,\mathscr{F})$. See the discussion near Definition 33.4 in [69].

$$= \Gamma(-\alpha)\int_{\mathbb{R}}\left[\left(1-\frac{(iz+\theta)x}{1+x\theta}\right)^{\alpha}-\left(1-\frac{\theta x}{1+x\theta}\right)^{\alpha}\right](1+x\theta)^{\alpha}R_{\theta}(\mathrm{d}x)+izb_0$$

$$= \Gamma(-\alpha)\int_{\mathbb{R}}\left[(1-izx)^{\alpha}-1\right]R_{\theta}(\mathrm{d}x)+izb_0,$$

where

$$R_{\theta}(A)=\int_{\mathbb{R}}1_A\left(\frac{x}{1-x\theta}\right)(1-x\theta)^{\alpha}R(\mathrm{d}x).$$

It is straightforward to check that

$$R(A)=\int_{\mathbb{R}}1_A\left(\frac{x}{1+x\theta}\right)(1+x\theta)^{\alpha}R_{\theta}(\mathrm{d}x)$$

and

$$\int_{\mathbb{R}}|x|^{\alpha}R_{\theta}(\mathrm{d}x)=\int_{\mathbb{R}}|x|^{\alpha}R(\mathrm{d}x)<\infty.$$

From Section 9.5 in [21] and the above discussion it follows that, under measure Q^{θ}, the process $\{X_t : t \geq 0\}$ is a Lévy process with $X_1 \sim TS^1_{\alpha}(R_{\theta}, b_{\theta})$ where

$$b_{\theta}=b_0+\int_{\mathbb{R}}\int_0^{\infty}\frac{x}{1+t^2x^2}t^{-\alpha}e^{-t}\mathrm{d}tR_{\theta}(\mathrm{d}x).$$

Note that, in the above, b_0 does not depend on θ. By arguments similar to the proof of Theorem 7.1 we get the following.

Theorem 7.2. *The equivalent probability measure Q^{θ} is a risk-neutral measure if and only if $R_{\theta}(x>1)=0$ and*

$$\Gamma(-\alpha)\int_{\mathbb{R}}[(1-x)^{\alpha}-1]R_{\theta}(\mathrm{d}x)+b_0=r. \tag{7.2}$$

Although $\Theta \neq \emptyset$, it is possible to have $\Theta = \{0\}$. In fact, this is the case for the distributions discussed in Section 6.2. When this holds we cannot use the Esscher transform. However, we can use the so-called asymmetric or bilateral Esscher transform. We begin by introducing some notation. Let

$$R^{+}(A)=R(A\cap(0,\infty)) \text{ and } R^{-}(A)=R(A\cap(-\infty,0)), \qquad A\in\mathfrak{B}(\mathbb{R})$$

and let

$$X_t^{+}=\sum_{s\in[0,t]}\Delta X_s 1_{\Delta X_s>0} \text{ and } X_t^{-}=\sum_{s\in[0,t]}\Delta X_s 1_{\Delta X_s<0}.$$

From Corollary 3.1 in [21] it follows that

$$X_t \stackrel{d}{=} X_t^+ + X_t^- + tb',$$

where $b' = b - b^+ - b^-$ with

$$b^- = \int_{(-\infty,0)} \int_0^\infty \frac{x}{1+t^2x^2} t^{-\alpha} e^{-t} \mathrm{d}t R(\mathrm{d}x)$$

and

$$b^+ = \int_{(0,\infty)} \int_0^\infty \frac{x}{1+t^2x^2} t^{-\alpha} e^{-t} \mathrm{d}t R(\mathrm{d}x).$$

Moreover, $\{X_t^- : t \geq 0\}$ and $\{X_t^+ : t \geq 0\}$ are independent Lévy processes[3] with $X_1^- \sim TS_\alpha^1(R^-, b^-)$ and $X_1^+ \sim TS_\alpha^1(R^+, b^+)$.

Set $\Theta^- = [-\eta_-^{-1}, \infty)$, $\Theta^+ = (-\infty, \eta_+^{-1}]$, and note that by Lemma 3.27 for any $\theta^- \in \Theta^-$ and any $\theta^+ \in \Theta^+$ we have

$$\mathrm{E}[e^{\theta^- X_1^-}] < \infty \text{ and } \mathrm{E}[e^{\theta^+ X_1^+}] < \infty.$$

For any $\theta^- \in \Theta^-$ and any $\theta^+ \in \Theta^+$ define a probability measure Q^{θ^-,θ^+} on $(\Omega, \mathscr{F})$ by the Radon-Nikodym derivative process

$$\frac{\mathrm{d}Q^{\theta^-,\theta^+}_{|\mathscr{F}_t}}{\mathrm{d}P_{|\mathscr{F}_t}} = \frac{e^{\theta^- X_t^-} e^{\theta^+ X_t^+}}{\mathrm{E}[e^{\theta^- X_t^-} e^{\theta^- X_t^-}]} = e^{\theta^- X_t^- + \theta^+ X_t^+ - t\Psi_\alpha^-(\theta^-) - t\Psi_\alpha^+(\theta^+)},$$

where

$$\Psi_\alpha^+(\theta^+) = \Gamma(-\alpha) \int_{(0,\infty)} [(1 - \theta^+ x)^\alpha - 1] R^+(\mathrm{d}x)$$

and

$$\Psi_\alpha^-(\theta^-) = \Gamma(-\alpha) \int_{(-\infty,0)} [(1 - \theta^- x)^\alpha - 1] R^-(\mathrm{d}x).$$

By arguments similar to the previous case we can show that under the equivalent measure Q^{θ^-,θ^+} the process $\{X_t : t \geq 0\}$ is a Lévy process with $X_1 \sim TS_\alpha^1(R_{\theta^-,\theta^+}, b_{\theta^-,\theta^+})$, where

---

[3] This follows from properties of Poisson random measures and the fact that the jump measure of a Lévy process is a Poisson random measure, see, e.g., Theorem 19.2 in [69] or Chapter 3 in [21].

$$R_{\theta^-,\theta^+}(A) = \int_{(-\infty,0)} 1_A\left(\frac{x}{1-x\theta^-}\right)(1-x\theta^-)^\alpha R(\mathrm{d}x) + \int_{(0,\infty)} 1_A\left(\frac{x}{1-x\theta^+}\right)(1-x\theta^+)^\alpha R(\mathrm{d}x)$$

and

$$b_{\theta^-,\theta^+} = b' + \int_{\mathbb{R}}\int_0^\infty \frac{x}{1+t^2x^2} t^{-\alpha} e^{-t} \mathrm{d}t R_{\theta^-,\theta^+}(\mathrm{d}x).$$

Arguments similar to the proof of Theorem 7.1 give the following.

Theorem 7.3. *The equivalent probability measure Q^{θ^-,θ^+} is a risk-neutral measure if and only if $R_{\theta^-,\theta^+}(x > 1) = 0$ and*

$$\Gamma(-\alpha)\int_{\mathbb{R}}[(1-x)^\alpha - 1]R_{\theta^-,\theta^+}(\mathrm{d}x) + b' = r. \tag{7.3}$$

It is straightforward to see that $R_{\theta^-,\theta^+}(x > 1) = 0$ always holds when $\theta^+ \leq -1$ and $\theta^- \geq 0$. In order to find an equivalent risk-neutral measure it suffices to find such values of θ^- and θ^+ that solve (7.3). For certain parametric classes of tempered stable distributions approaches for solving such equations and for finding other equivalent risk-neutral measures are given in [46, 60], and [50].

7.2 Mobility Models

For many applications it is important to understand the movement of an animal or a person through some terrain. A model for this is called a mobility model. In an ecological context it may represent an animal foraging for food (see, e.g., [43] and the references therein). In an anthropological context it may represent the movement of hunter-gatherers (see [61]). In other applications it may represent the movement of a person in his or her daily life. This last situation is particularly important for computer science. This is due to the fact that many people carry cell phones and other mobile devices, and, in order to develop and evaluate routing protocols for these devices, it is imperative to be able to simulate the movements of humans in a realistic way (see [16, 64], and the references therein).

Interestingly, in all of the situations discussed above, similar models tend to appear. We will show that these models are well approximated by TW_α^p distributions, which were introduced in Section 6.1. We begin by discussing the movement of "the walker," which may represent a human or an animal. Assume that we observe the location of the walker at fixed time increments $\Delta > 0$. Let $\{X_n : n = 0, 1, \ldots\}$ be a discrete time stochastic process on $\mathbb{R}^2$ such that X_n is the location of the walker at time $n\Delta$. Let $Z_n = X_n - X_{n-1}$ be the increment process. If the walker did not stay in the same place during the ith time interval, then we can write $Z_i = \frac{Z_i}{|Z_i|}|Z_i|$, where $|Z_i|$ represents the magnitude of the displacement and $\frac{Z_i}{|Z_i|}$ represents the direction of travel. We focus on modeling the distribution of $|Z_i|$. Models for the distribution of $\frac{Z_i}{|Z_i|}$ and for the case when $Z_i = 0$ are discussed in [16].

A common model for $|Z_i|$ is called a Lévy walk. Here it is assumed that the magnitudes of displacement, i.e. the $|Z_i|$'s, are iid random variables having a Pareto distribution, i.e. having a probability density given by

$$f(x) = \begin{cases} \alpha\delta^{\alpha}x^{-\alpha-1} & x > \delta \\ 0 & \text{otherwise} \end{cases}, \tag{7.4}$$

where $\alpha \in (0, 1)$ and $\delta > 0$ are parameters. One can allow α to be any positive number, but most empirical data suggests $\alpha \in (0, 1)$, see [26, 61, 64], and the references therein. In practice, however, movement is more complicated and the Pareto distribution is only valid for large values of x. For this reason we only assume that the density satisfies $f(x) = 0$ for $x \leq 0$ and

$$f(x) \sim kx^{-\alpha-1} \text{ as } x \to \infty. \tag{7.5}$$

for some $k > 0$ and $\alpha \in (0, 1)$.

It turns out that the actual distribution of $|Z_i|$ does not matter if we model the lengths of entire flights at once. A flight is a part of the walk where the walker keeps going in the same direction. Assume, for instance, that after the walker starts walking at time 0, the first time that he or she stops or changes direction is at time $N\Delta$. This means that $|X_N| = |Z_1 + Z_2 + \cdots + Z_N| = \sum_{i=1}^{N} |Z_i|$, where the second equality follows because all of the steps are in the same direction. Since a person is likely to walk in the same direction for a relatively long time, N is likely to be quite large and we can approximate the distribution of $|X_N|$ by its asymptotic distribution. This asymptotic distribution is described by the theorem below, which is a version of the central limit theorem for infinite variance distributions, see, e.g., [23] for details.

Theorem 7.4. *If $\alpha \in (0, 1)$, then*

$$n^{-1/\alpha} \sum_{i=1}^{n} |Z_i| \xrightarrow{d} Y,$$

where Y has a fully right skewed α-stable distribution with Laplace transform

$$\mathrm{E}[e^{-zY}] = e^{-c|z|^{\alpha}}, \text{ for } z \geq 0.$$

Here the parameter $c > 0$ is a constant depending on the distribution of $|Z_i|$.

This suggests that flight lengths should be well modeled by such α-stable distributions. In other words, it seems that one merely needs to fit this two parameter model to data. However, the tails of these distributions are too heavy. In practice, there are various geographic and physical limitations that prevent flight lengths from getting too big. In fact, empirical data suggests that (7.5) holds for large, but not too large values of x, see [26, 61, 64], and the references therein. This has lead to the development of tempered Lévy walks.

A tempered Lévy walk assumes that the density of $|Z_i|$ satisfies $f(x) = 0$ for $x \leq 0$ and

$$f(x) \sim ke^{-(x/\ell)^p} x^{-\alpha-1} \text{ as } x \to \infty, \tag{7.6}$$

for $\alpha \in (0,1)$, $k > 0$, $\ell > 0$, and $p > 0$. It is commonly assumed that $p = 1$, but we will not do so here. If ℓ is very large, this means that, for medium and somewhat large values of x, we have $f(x) \approx kx^{-\alpha-1}$, but for very large values of x we start to feel the exponential function and the tails ultimately decay exponentially fast. We will give a limit theorem, which will show that, in this case, the sum $|X_N| = \sum_{i=1}^N |Z_i|$ is well approximated by a tempered stable distribution. Toward this end, we introduce the notation $Z_i(\ell) = Z_i$ to emphasize the dependence of the distribution of Z_i on the parameter ℓ.

Theorem 7.5. *If $\alpha \in (0,1)$ and $\ell_n \to \infty$ such that $n^{-1/\alpha}\ell_n \to \ell \in (0,\infty)$, then*

$$n^{-1/\alpha} \sum_{i=1}^{N} |Z_i(\ell_n)| \xrightarrow{d} Y \text{ as } n \to \infty,$$

where $Y \sim TW_\alpha^p(k,\ell,b)$ and

$$b = k \int_0^\infty \frac{x}{1+x^2} e^{-(x/\ell)^p} x^{-1-\alpha} \mathrm{d}x.$$

We postpone the proof of Theorem 7.5 to Section 7.3, where it will follow from a more general result. The value of b in the above ensures that the support of the distribution of Y is $[0,\infty)$, see Remark 3.6. Note that, in the theorem, we assume that $\ell_n \to \infty$, but in practice we generally have a fixed parameter ℓ that is not changing. In this case, we interpret the theorem as follows. If n is large, but $n^{-1/\alpha}\ell$ is "medium sized," we can approximate the distribution of the sum by the given TW_α^p distribution. This suggests that such distributions should provide good models for flight lengths. In fact [16] showed that they provide a good fit to empirical data.

7.3 How Do Tempered Stable Distributions Appear in Applications?

In Section 7.2 we saw a theoretical explanation for why TW_α^p distributions with $\alpha \in (0,1)$ provide good models for flight lengths in mobility models. In this section we give more general results, which aim to explain the theoretical mechanism by which other tempered stable distributions appear in applications. We begin with some definitions.

Let μ be a probability measure on $\mathbb{R}$ in the domain of attraction of some infinite variance α-stable law. For simplicity assume that $\mu((-\infty, 0)) = 0$. For $p, \ell \in (0, \infty)$ define a new probability measure by

$$\mu_p^{(\ell)}(\mathrm{d}x) = c_\ell e^{-(x/\ell)^p} \mu(\mathrm{d}x), \tag{7.7}$$

where

$$c_\ell = \left[\int_{[0,\infty)} e^{-(x/\ell)^p} \mu(\mathrm{d}x)\right]^{-1}$$

is a normalizing constant. We call this **exponential tempering** of μ. Clearly, $\mu_p^{(\ell)}$ has all moments finite and belongs to the domain of attraction of the Gaussian. However, if ℓ is very large, then $\mu_p^{(\ell)}$ will be similar to μ in some central region, but the chance of a very large value will be "tempered."

Since μ belongs to the domain of attraction of an α-stable distribution with $\alpha \in (0, 2)$ there exists a function $L \in RV_0^\infty$ with

$$\mu(x > t) = t^{-\alpha} L(t).$$

Define

$$V(t) = t^\alpha / L(t) \text{ and } a_t = 1/V^{\leftarrow}(t) \tag{7.8}$$

and let $X_1, X_2, \cdots \stackrel{\text{iid}}{\sim} \mu$. The central limit theorem for infinite variance distributions (see, e.g., [23]) implies that there exists a sequence $\{\eta_n\}$ in $\mathbb{R}$ with

$$a_n \sum_{i=1}^n (X_i - \eta_n) \stackrel{d}{\to} R_{\alpha,\infty},$$

where $R_{\alpha,\infty} \sim ID(0, M_\infty, 0)$ with

$$M_\infty(\mathrm{d}x) = \alpha x^{-1-\alpha} 1_{[x>0]} \mathrm{d}x \tag{7.9}$$

is an α-stable random variable.

Now fix $p > 0$ and let $\{\ell_n\}$ be a sequence of positive real numbers such that

$$\ell_n \to \infty.$$

Define $\mu_p^{(\ell_n)}$ as in (7.7) and set

$$S_n(\ell_n) = \sum_{i=1}^n X_i(\ell_n), \tag{7.10}$$

where $X_1(\ell_n), \ldots, X_n(\ell_n) \stackrel{\text{iid}}{\sim} \mu_p^{(\ell_n)}$. Theorem 5 in [34] gives the following.[4]

[4] A version of this result also appears in [33]. It should be noted that we are using a slightly different parametrization than the one used in [33] and [34].

Theorem 7.6. *1. If* $a_n\ell_n \to \ell \in (0,\infty)$, *then*

$$a_n S_n(\ell_n) - \kappa_n \xrightarrow{d} R_{\alpha,\ell} + \eta_{\alpha,\ell}$$

where $R_{\alpha,\ell} \sim TW_\alpha^p(\alpha,\ell,0)$,

$$\kappa_n = c_{\ell_n} n a_n \int_{[0,1/a_n]} x e^{-(x/\ell_n)^p} \mu(\mathrm{d}x), \text{ and}$$

$$\eta_{\alpha,\ell} = \int_0^\infty x\left(\frac{1}{1+x^2} - 1(x<1)\right)\alpha e^{-(x/\ell)^p} x^{-1-\alpha}\mathrm{d}x.$$

2. *If* $a_n\ell_n \to \infty$, *then*

$$a_n S_n(\ell_n) - \kappa_n \xrightarrow{d} R_{\alpha,\infty} + \eta_{\alpha,\infty}$$

where

$$\kappa_n = c_{\ell_n} n a_n \int_{[0,1/a_n]} x e^{-(x/\ell_n)^p} \mu(\mathrm{d}x),$$

$$\eta_{\alpha,\infty} = \int_0^\infty x\left(\frac{1}{1+x^2} - 1(x<1)\right)\alpha x^{-1-\alpha}\mathrm{d}x,$$

and $R_{\alpha,\infty}$ *has the* α*-stable distribution* $ID(0, M_\infty, 0)$ *with* M_∞ *as in* (7.9).

3. *If* $a_n\ell_n \to 0$, *then*

$$b_n S_n(\ell_n) - \kappa_n' \xrightarrow{d} N(0,\sigma^2),$$

where

$$\sigma^2 = \frac{\alpha}{p}\Gamma\left(\frac{2-\alpha}{p}\right),$$

$$b_n = n^{-1/2}\ell_n^{-1}\sqrt{V(\ell_n)} = n^{-1/2}\ell_n^{-(2-\alpha)/2}\left[L(\ell_n)\right]^{-1/2}, \text{ and}$$

$$\kappa_n' = c_{\ell_n} n b_n \int_{[0,1/b_n]} x e^{-(x/\ell_n)^p} \mu(\mathrm{d}x).$$

In practice, for most applications, the parameter ℓ_n is not actually approaching infinity. Instead it is some fixed but (very) large constant ℓ. We can write $a_n\ell = [n^{-1}\ell^\alpha]^{1/\alpha}L'(n)$ for some $L' \in RV_0^\infty$. If $S_n(\ell)$ is the sum of n iid random variables with distribution $\mu_p^{(\ell)}$, then Theorem 7.6 can be interpreted as follows. When n is on the order of ℓ^α, the distribution is close to a TW_α^p distribution. However, once n is much larger than ℓ^α the distribution of $S_n(\ell)$ is well approximated by the Gaussian.

A constant that determines when such regimes occur is called the **natural scale** and was introduced in [35]. Thus ℓ^α is the natural scale for this model.[5]

The above discussion provides an explanation for how TW_α^p distributions with $\alpha \in (0, 2)$ appear in applications. Specifically, in applications where we model sums of exponentially tempered random variables. As discussed in Sections 7.1 and 7.2 such models are reasonable when modeling flight lengths and log-returns.

Now recall that TW_α^p distributions are the building blocks from which all other tempered stable distributions are constructed. Specifically, every p-tempered α-stable distribution on $\mathbb{R}^d$ is the limit, in distribution, of a sequence of linear combinations of independent random variables in the class TW_α^p, see Theorem 4.18 and Section 6.1. This gives an explanation for how more general tempered stable distributions appear in application. Alternate, but related, explanations for the appearance of tempered stable distributions are given in [35] and [19]. We conclude this section by using Theorem 7.6 to prove Theorem 7.5.

Proof (Proof of Theorem 7.5). In this case $\alpha \in (0, 1)$ and $\mu(\mathrm{d}x) = m(x)\mathrm{d}x$, where $m(x) = 0$ for $x < 0$ and $m(x) \sim kx^{-1-\alpha}$ as $x \to \infty$. This means that

$$m(x) = \alpha x^{-1-\alpha} L(x),$$

where $L \in RV_0^\infty$ and $\lim_{x\to\infty} L(x) = k/\alpha$. By Karamata's Theorem (Theorem 2.7) it follows that

$$\mu(x > t) = \alpha \int_t^\infty x^{-1-\alpha} L(x)\mathrm{d}x \sim t^{-\alpha} L(t) \text{ as } t \to \infty.$$

Define a_n by (7.8) and note that by Proposition 2.6

$$a_n \sim (\alpha/k)^{1/\alpha} n^{-1/\alpha}.$$

We have $n^{-1/\alpha}\ell_n \to \ell$ and hence $a_n \ell_n \to (\alpha/k)^{1/\alpha}\ell =: \ell'$. Now define $S_n(\ell_n)$ by (7.10) and note that the first part of Theorem 7.6 implies that

$$a_n S_n(\ell_n) - \kappa_n \stackrel{d}{\to} R_{\alpha,\ell'} + \eta_{\alpha,\ell'}.$$

We will show that $\lim_{n\to\infty} \kappa_n = \alpha \int_0^1 x^{-\alpha} e^{-(x/\ell')^p} \mathrm{d}x$. Assume, for the moment, that we already have this result. Combining it with Slutsky's Theorem gives

$$n^{-1/\alpha} S_n(\ell_n) \stackrel{d}{\to} \left(\frac{k}{\alpha}\right)^{1/\alpha} \left(R_{\alpha,\ell'} + \eta_{\alpha,\ell'} + \alpha \int_0^1 x^{-\alpha} e^{-(x/\ell')^p} \mathrm{d}x\right) =: R'_{\alpha,\ell'}.$$

[5] It should be noted that [34] reported a slightly different natural scale. This is due to that fact that a different parametrization was used there.

From here note that

$$\eta_{\alpha,\ell'} + \alpha \int_0^1 x^{-\alpha} e^{-(x/\ell')^p} \mathrm{d}x = \int_0^\infty \frac{x}{1+x^2} \alpha e^{-(x/\ell')^p} x^{-1-\alpha} \mathrm{d}x$$

and set $k' = (k/\alpha)^{1/\alpha}$. The characteristic function of $R'_{\alpha,\ell'}$ is given by $e^{C(z)}$, where

$$\begin{aligned}
C(z) &= \alpha \int_0^\infty \left(e^{ixk'z} - 1\right) e^{-(x/\ell')^p} x^{-1-\alpha} \mathrm{d}x \\
&= \alpha (k')^\alpha \int_0^\infty \left(e^{ixz} - 1\right) e^{-(x/(k'\ell'))^p} x^{-1-\alpha} \mathrm{d}x \\
&= k \int_0^\infty \left(e^{ixz} - 1\right) e^{-(x/\ell)^p} x^{-1-\alpha} \mathrm{d}x,
\end{aligned}$$

which is the characteristic function of $TW^p_\alpha(k, \ell, b)$, where

$$b = k \int_0^\infty \frac{x}{1+x^2} e^{-(x/\ell)^p} x^{-1-\alpha} \mathrm{d}x.$$

It remains to verify the limit of κ_n. Toward this end fix $\epsilon \in (0,1)$ and note that there exists a $K_\epsilon > 0$ such that for all $x > K_\epsilon$ we have

$$(1-\epsilon)\frac{k}{\alpha} \le L(x) \le (1+\epsilon)\frac{k}{\alpha}.$$

It follows that for large enough n

$$\begin{aligned}
(1-\epsilon) k c_{\ell_n} n a_n \int_{K_\epsilon}^{1/a_n} x^{-\alpha} e^{-(x/\ell_n)^p} \mathrm{d}x &\le \alpha c_{\ell_n} n a_n \int_{K_\epsilon}^{1/a_n} x^{-\alpha} e^{-(x/\ell_n)^p} L(x) \mathrm{d}x \\
&\le (1+\epsilon) k c_{\ell_n} n a_n \int_{K_\epsilon}^{1/a_n} x^{-\alpha} e^{-(x/\ell_n)^p} \mathrm{d}x.
\end{aligned}$$

By change of variables, dominated convergence, and the facts that $a_n \ell_n \to \ell'$, $c_{\ell_n} \to 1$, and $n a_n^\alpha \to \alpha/k$ we have

$$\begin{aligned}
k c_{\ell_n} n a_n \int_{K_\epsilon}^{1/a_n} x^{-\alpha} e^{-(x/\ell_n)^p} \mathrm{d}x &= k c_{\ell_n} n a_n^\alpha \int_{K_\epsilon a_n}^1 x^{-\alpha} e^{-[x/(a_n\ell_n)]^p} \mathrm{d}x \\
&\sim \alpha \int_0^1 x^{-\alpha} e^{-(x/\ell')^p} \mathrm{d}x,
\end{aligned}$$

where the integral is finite since $\alpha \in (0,1)$. Further,

$$0 \leq \alpha c_{\ell_n} n a_n \int_0^{K_\epsilon} x^{-\alpha} e^{-(x/\ell_n)^p} L(x) \mathrm{d}x \leq \alpha c_{\ell_n} n a_n \int_0^{K_\epsilon} x^{-\alpha} L(x) \mathrm{d}x$$
$$\sim \alpha(\alpha/k)^{1/\alpha} n^{1-1/\alpha} \int_0^{K_\epsilon} x^{-\alpha} L(x) \mathrm{d}x \to 0,$$

as $n \to \infty$. Note that the integral is finite since $m(x)$ is a probability density. Putting everything together gives

$$\limsup_{n\to\infty} \kappa_n \leq \lim_{\epsilon\downarrow 0} \lim_{n\to\infty} \alpha c_{\ell_n} n a_n \int_0^{K_\epsilon} x^{-\alpha} e^{-(x/\ell_n)^p} L(x) \mathrm{d}x$$
$$+ \lim_{\epsilon\downarrow 0} \lim_{n\to\infty} (1+\epsilon) k c_{\ell_n} n a_n \int_{K_\epsilon}^{1/a_n} x^{-\alpha} e^{-(x/\ell_n)^p} \mathrm{d}x$$
$$= \alpha \int_0^1 x^{-\alpha} e^{-(x/\ell')^p} \mathrm{d}x$$

and similarly

$$\liminf_{n\to\infty} \kappa_n \geq \alpha \int_0^1 x^{-\alpha} e^{-(x/\ell')^p} \mathrm{d}x,$$

which gives the result. □

Chapter 8
Epilogue

In this brief we discussed many properties of the class of p-tempered α-stable distributions and their associated Lévy processes. Our discussion encompassed much of what is known about these models and filled in many gaps in the literature. However, the goal of a brief is to, well, be brief, and as such there are a number of topics that we were unable to include. In this epilogue we discuss several such topics and give references to the literature.

One topic of interest is the study of stochastic integral representations of tempered stable distributions. We refer the reader to [6, 40, 51], and [32].

Another topic is the study of the class of normal tempered stable distributions. These models are obtained as follows. Let $\{T_t : t \geq 0\}$ be a tempered stable subordinator (see Proposition 3.24 for a characterization) and let $\{W_t : t \geq 0\}$ be a multivariate Brownian motion independent of $\{T_t : t \geq 0\}$. The time changed Brownian motion

$$\{W_{T_t} : t \geq 0\}$$

is, by Theorem 30.1 in [69], a Lévy process. The marginal distributions of this process are called normal tempered stable distributions, and are studied in, e.g., [7, 28, 31], and the references therein.

A third topic involves defining tempered stable processes, i.e. stochastic processes with tempered stable finite-dimensional distributions. The best known processes of this type are the tempered stable *Lévy* processes that we focused on in this brief. While these are important and understanding their behavior helps to understand the behavior of more general tempered stable processes, they have a fairly simple dependence structure, which is not adequate for many applications.

One way to get models with more intricate dependence structures is to consider stochastic integration. Specifically, consider the processes $X_t = \int_0^\infty f(s,t)\mathrm{d}Y_s$, where $\{Y_s : s \geq 0\}$ is a Lévy process and $f(s,t)$ is some deterministic function for which the integral exists, see [70] for definitions and information about such

M. Grabchak, *Tempered Stable Distributions*, SpringerBriefs
in Mathematics, DOI 10.1007/978-3-319-24927-8_8

integrals. One can either assume that $\{Y_s : s \geq 0\}$ is a tempered stable Lévy process or that it is another Lévy process, but one which, nevertheless, leads to a tempered stable process. Examples of such models include tempered stable Ornstein-Uhlenbeck-type processes studied in, e.g., [76] and [31], and fractional tempered stable motion introduced in [38].

For certain applications, it is of interest to obtain normal tempered stable process. This can be done by time changing a Brownian motion by a positive and increasing tempered stable process. An example of this based on a modification of a tempered stable Ornstein-Uhlenbeck-type process is given in [44] and an example based on fractional tempered stable motion is given in [45].

Note that all of the tempered stable and normal tempered stable processes discussed above start with a Lévy process. This is not necessary, but we are only aware of one example from the literature where this is not the case. The weighted tempered stable moving average process introduced in [25] is a tempered stable process defined in terms of integration not with respect to a Lévy processes, but with respect to a more general independently scattered tempered stable random measure. It may be of interest to define other processes in this manner.

Another approach is to see what can be done in the following general setting. Let T be an index set and let $\{X_t : t \in T\}$ be a collection of random variables such that for any $t_1, t_2, \ldots, t_d \in T$ the vector $(X_{t_1}, X_{t_2}, \ldots, X_{t_d})$ follows a tempered stable distribution on $\mathbb{R}^d$. This construction mimics that of general infinitely divisible processes introduced in [53]. We are aware of no work in this direction, but it seems that interesting results can be proved even at this level of generality.

References

1. O. O. Aalen (1992). Modelling heterogeneity in survival analysis by the compound Poisson distribution. *The Annals of Applied Probability*, 2(4): 951–972.
2. M. Abramowitz and I. A. Stegun (1972). *Handbook of Mathematical Functions* 10th ed. Dover Publications, New York.
3. T. Aoyama, M. Maejima, and J. Rosiński (2008). A subclass of type G selfdecomposable distributions on $\mathbb{R}^d$. *Journal of Theoretical Probability*, 21(1):14–34.
4. B. Baeumer and M. M. Meerschaert (2010). Tempered stable Lévy motion and transient super-diffusion. *Journal of Computational and Applied Mathematics*, 233:2438–2448.
5. A. A. Balkema (1973). *Monotone Transformations and Limit Laws*. Mathematisch Centrum, Amsterdam.
6. O. E. Barndorff-Nielsen, M. Maejima, and K. Sato (2006). Some classes of multivariate infinitely divisible distributions admitting stochastic integral representations. *Bernoulli*, 12(1):1–33.
7. O. E. Barndorff-Nielsen and N. Shephard (2002). Normal modified stable processes. *Theory of Probability and Mathematical Statistics*, 65:1–20.
8. H. Bauer (1981). *Probability Theory and Elements of Measure Theory*, 2nd English Ed. Academic Press, London. Translated by R. B. Burckel.
9. M. L. Bianchi, S. T. Rachev, Y. S. Kim, and F. J. Fabozzi (2011). Tempered infinitely divisible distributions and processes. *Theory of Probability and Its Applications*, 55(1):2–26.
10. P. Billingsley (1995). *Probability and Measure*, 3rd Ed. Wiley, New York.
11. N. H. Bingham, C. M. Goldie, and J. L. Teugels (1987). *Regular Variation*. Encyclopedia of Mathematics And Its Applications. Cambridge University Press, Cambridge.
12. R. M. Blumenthal and R. K. Getoor (1961). Sample functions of stochastic processes with stationary independent increments. *Journal of Mathematics and Mechanics*, 10:493–516.
13. V. I. Bogachev (2007). *Measure Theory Volume II*. Springer-Verlag, Berlin.
14. R. Bruno, L. Sorriso-Valvo, V. Carbone, and B. Bavassano (2004). A possible truncated-Lévy-flight statistics recovered from interplanetary solar-wind velocity and magnetic-field fluctuations. *Europhysics Letters*, 66(1): 146–152.
15. S. I. Boyarchenko and S. Levendorskiĭ (2000). Option pricing for truncated Lévy processes. *International Journal of Theoretical and Applied Finance*, 3(3): 549–552.
16. L. Cao and M. Grabchak (2014). Smoothly truncated Lévy walks: Toward a realistic mobility model. *IPCCC '14: Proceedings of the 33rd International Performance Computing and Communications Conference*.
17. P. Carr, H. Geman, D. B. Madan, and M. Yor (2002). The fine structure of asset returns: An empirical investigation. *Journal of Business*, 75(2): 305–332.

M. Grabchak, *Tempered Stable Distributions*, SpringerBriefs
in Mathematics, DOI 10.1007/978-3-319-24927-8

18. D. S. Carter (1958). L'Hospital's rule for complex-valued functions. *The American Mathematical Monthly*, 65(4), 264–266.
19. A. Chakrabarty and M. M. Meerschaert (2011). Tempered stable laws as random walk limits. *Statistics & Probability Letters*, 81(8):989–997.
20. J. M. Chambers, C. L. Mallows and B. W. Stuck (1976). A method for simulating stable random variables. *Journal of the American Statistical Association*, 71(354), 340–344.
21. R. Cont and P. Tankov (2004). *Financial Modeling With Jump Processes*. Chapman & Hall, Boca Raton.
22. R. M. Dudley and R. Norvaiša (1998). *An Introduction to p-Variation and Young Integrals with emphasis on sample functions of stochastic processes*. Number 1 in MaPhySto Lecture Notes. Center for Mathematical Physics and Stochastics.
23. W. Feller (1971). *An Introduction to Probability Theory and Its Applications Volume II*, 2nd Ed. John Wiley & Sons, Inc., New York.
24. B. Fristedt and S. J. Taylor (1973). Strong variation for the sample functions of a stable process. *Duke Mathematical Journal*, 40(2):259–278.
25. J. L. P. Garmendia (2008). On weighted tempered moving averages processes. *Stochastic Models*, 24(Suppl1):227–245.
26. M. C. González, C. A. Hidalgo, and A. L. Barabási (2008). Understanding individual human mobility patterns. *Nature*, 453(7169):779–782.
27. M. Grabchak (2012). On a new class of tempered stable distributions: Moments and regular variation. *Journal of Applied Probability*, 49(4):1015–1035.
28. M. Grabchak (2014). Does value-at-risk encourage diversification when losses follow tempered stable or more general Lévy processes? *Annals of Finance*, 10(4):553–568.
29. M. Grabchak (2015a). A simple condition for the multivariate CLT and the attraction to the Gaussian of Lévy processes at long and short times. To appear in *Communications in Statistics-Theory and Methods*. Preprint available at http://arxiv.org/abs/1306.1885.
30. M. Grabchak (2015b). Inversions of Lévy measures and the relation between long and short time behavior of Lévy processes. *Journal of Theoretical Probability*, 28(1):184–197.
31. M. Grabchak (2015c). On the consistency of the MLE for Ornstein-Uhlenbeck and other selfdecomposable processes. *Statistical Inference for Stochastic Processes*, DOI 10.1007/s11203-015-9118-9.
32. M. Grabchak (2015d). Three upsilon transforms related to tempered stable distributions. *Electronic Communication in Probability*, 20(82):1–10.
33. M. Grabchak and S. Molchanov (2013). Limit theorems and phase transitions for two models of summation of iid random variables depending on parameters. *Doklady Mathematics*, 88(1):431–434.
34. M. Grabchak and S. Molchanov (2015). Limit theorems and phase transitions for two models of summation of i.i.d. random variables with a parameter. *Theory of Probability and Its Applications*, 59(2):222–243.
35. M. Grabchak and G. Samorodnitsky (2010). Do financial returns have finite or infinite variance? A paradox and an explanation. *Quantitative Finance*, 10(8):883–893.
36. P. S. Griffin, R. A. Maller, and D. Roberts (2013). Finite time ruin probabilities for tempered stable insurance risk processes. *Insurance: Mathematics and Economics*, 53(2): 478–489.
37. D. Hainaut and P. Devolders (2008). Mortality modelling with Lévy processes. *Insurance: Mathematics and Economics*, 42(1):409–418.
38. C. Houdré and R. Kawai (2006) On fractional tempered stable motion. *Stochastic Processes and Their Applications*, 116(8):1161–1184.
39. P. Hougaard (1986). Survival models for heterogeneous populations derived from stable distributions. *Biometrika*, 73(2): 387–396.
40. Z. J. Jurek (2007). Random integral representations for free-infinitely divisible and tempered stable distributions. *Statistics & Probability Letters*, 77(4):417–425.
41. O. Kallenberg (2002). *Foundations of Modern Probability* 2nd ed. Springer, New York.
42. R. Kawai and H. Masuda (2011). On simulation of tempered stable random variates. *Journal of Computational and Applied Mathematics*, 235:2873–2887.

43. R. Kawai and S. Petrovskii (2012). Multi-scale properties of random walk models of animal movement: Lessons from statistical inference. *Proceedings of the Royal Society A*, 468(2141): 1428–1451.
44. A. D. J. Kerss, N. N. Leonenko, and A. Sikorskii (2014). Risky asset models with tempered stable fractal activity time. *Stochastic Analysis and Applications* , 32(4), 642–663.
45. Y. S. Kim (2012). The fractional multivariate normal tempered stable process. *Applied Mathematics Letters*, 25(12), 2396–2401.
46. Y. S. Kim, S. T. Rachev, M. L. Bianchi, and F. J. Fabozzi (2009). A new tempered stable distribution and its application to finance. In G. Bol, S. T. Rachev, and R. Würth (eds.), *Risk Assessment: Decisions in Banking and Finance*. Physica-Verlag, Springer, Heidelberg pg. 77–108.
47. Y. S. Kim, S. T. Rachev, M. L. Bianchi, and F. J. Fabozzi (2010). Computing VaR and AVaR in infinitely divisible distributions. *Probability and Mathematical Statistics*, 30(2), 223–245.
48. I. Koponen (1995). Analytic approach to the problem of convergence of truncated Lévy flights towards the Gaussian stochastic process. *Physical Review E*, 52(1):1197–1199.
49. U. Küchler and S. Tappe (2013). Tempered stable distributions and processes. *Stochastic Processes and their Applications*, 123(12):4256–4293.
50. U. Küchler and S. Tappe (2014). Exponential stock models driven by tempered stable processes. *Journal of Econometrics*, 181(1):53–63.
51. M. Maejima and G. Nakahara (2009). A note on new classes of infinitely divisible distributions on $\mathbb{R}^d$. *Electronic Communications in Probability*, 14:358–371.
52. R. N. Mantegna and H. E. Stanley (1994). Stochastic process with ultraslow convergence to a Gaussian: The truncated Lévy flight. *Physical Review Letters*, 73(22):2946–2949.
53. G. Maruyama (1970). Infinitely divisible processes. *Theory of Probability and Its Applications*, 15(1):1–22.
54. M. M. Meerschaert and H. Scheffler (2001). *Limit Distributions for Sums of Independent Random Vectors: Heavy Tails in Theory and Practice*. John Wiley & Sons, New York.
55. M. M. Meerschaert, Y. Zhang, and B. Baeumer (2008). Tempered anomalous diffusion in heterogeneous systems. *Geophysical Research Letters*, 35.
56. I. Monroe (1972). On the γ-variation of processes with stationary independent increments. *Annals of Mathematical Statistics*, 43(4):1213–1220.
57. J. P. Nolan (1997). Numerical Calculation of Stable Densities and Distributions. *Communications in Statistics–Stochastic Models*, 13(4): 759–774.
58. E. A. Novikov (1994). Infinitely divisible distributions in turbulence. *Physical Review E*, 50(5):R3303–R3305.
59. K. J. Palmer, M. S. Ridout, and B. J. T. Morgan (2008). Modelling Cell Generation times by using the tempered stable distribution. *Journal of the Royal Statistical Society Series C: Applied Statistics*, 57(4): 379–397.
60. S. T. Rachev, Y. S. Kim, M. L. Bianchi, and F. J. Fabozzi (2011). *Financial Models with Levy Processes and Volatility Clustering*. John Wiley & Sons Ltd.
61. D. A. Raichlen, B. M. Wood, A. D. Gordon, A. Z. P. Mabulla, F. W. Marlowe, and H. Pontzer (2014). Evidence of Lévy walk foraging patterns in human hunter–gatherers. *Proceedings of the National Academy of Sciences of the United States of America*, 111(2):728–733.
62. S. I. Resnick (1987). *Extreme Values, Regular Variation, and Point Processes*. Springer-Verlag, New York.
63. S. I. Resnick (2007). *Heavy-Tail Phenomena: Probabilistic and Statistical Modeling*. Springer, New York.
64. I. Rhee, M. Shin, S. Hong, K. Lee, S. J. Kim, and S. Chong (2011). On the levy-walk nature of human mobility: Do humans walk like monkeys? *IEEE/ACM Transaction on Networking*, 19(3):630–643.
65. J. Rosiński (2007). Tempering stable processes. *Stochastic Processes and their Applications*, 117(6):677–707.
66. J. Rosiński and J. L. Sinclair (2010). Generalized tempered stable processes. *Banach Center Publications*, 90:153–170.

67. E. L. Rvačeva (1962). On domains of attraction of multi-dimensional distributions. In *Selected Translations in Mathematical Statistics and Probability Vol. 2*, pg. 183–205. American Mathematical Society, Providence. Translated by S. G. Ghurye.
68. G. Samorodnitsky and M. S. Taqqu (1994). *Stable Non-Gaussian Random Processes: Stochastic Models with Infinite Variance*. Chapman & Hall, New York.
69. K. Sato (1999). *Lévy Processes and Infinitely Divisible Distributions*. Cambridge University Press, Cambridge.
70. K. Sato (2006). Additive processes and stochastic integrals. *Illinois Journal of Mathematics*, 50(4): 825–851.
71. V. Seshadri (1993). *The Inverse Gaussian Distribution: A Case Study in Exponential Families*. Oxford University Press, Oxford.
72. R. L. Schilling, R. Song, and Z. Vondraček (2012). *Bernstein Functions: Theory and Applications*. DeGruyter, Berlin.
73. P. J. Smith (1995). A recursive formulation of the old problem of obtaining moments from cumulants and vice versa. *The American Statistician*, 49(2): 217–218.
74. G. Terdik and T. Gyires (2009a). Lévy flights and fractal modeling of internet traffic. *IEEE/ACM Transactions on Networking*, 17(1):120–129.
75. G. Terdik and T. Gyires (2009b). Does the internet still demonstrate fractal nature? In *ICN '09: Eighth International Conference on Networks*, pg. 30–34.
76. G. Terdik and W. A. Woyczyński (2006). Rosiński Measures for tempered stable and related Ornstien-Uhlenbeck processes. *Probability and Mathematical Statistics*, 26(2): 213–243.
77. M. C. K. Tweedie (1984). An index which distinguishes between some important exponential families. In J. K. Ghosh and J. Roy (eds.), *Statistics: Applications and New Directions. Proceedings of the Indian Statistical Institute Golden Jubilee International Conference*. Indian Statistical Institute, Calcutta, pg. 579–604.
78. V. V. Uchaikin and V. M. Zolotarev (1999). *Chance and Stability: Stable Distributions and their Applications*. VSP BV, Utrecht.

Index

M. Grabchak, *Tempered Stable Distributions*, SpringerBriefs in Mathematics, DOI 10.1007/978-3-319-24927-8

Printed in the United States
By Bookmasters